Studies on Contemporary Europe

Edited by
PROFESSOR A. S. MILWARD
University of Manchester Institute of Science and Technology

2

Agriculture and the
European Community

Studies on Contemporary Europe

1   BUDGETARY POLITICS: THE FINANCES OF THE
     EUROPEAN COMMUNITIES
     by Helen Wallace

Agriculture and the European Community

JOHN S. MARSH
Professor of Agricultural Economics

PAMELA J. SWANNEY
Agricultural Economist

School of Agriculture, Aberdeen

University Association for Contemporary European Studies
George Allen & Unwin

First published in 1980

GEORGE ALLEN & UNWIN LTD
40 Museum Street, London WC1A 1LU

© University Association for Contemporary European Studies, 1980

British Library Cataloguing in Publication Data

Marsh, John Stanley
    Agriculture and the European Community.
    (Studies on contemporary Europe; 2).
    1. Agriculture – Economic aspects –
European Economic Community countries
    I. Title   II. Swanney, Pamela J   III. Series
    338.1′81   HD1920.5.Z8   80–40637

    ISBN 0–04–338092–1
    ISBN 0–04–338093–X Pbk

Set in 11 on 12 point Plantin by Computacomp (UK) Ltd, Fort William, Scotland and printed in Great Britain
by Billing and Sons Ltd., Guildford, London and Worcester

Contents

Editor's Preface

The University Association for Contemporary European Studies (UACES) exists to promote the study of contemporary European society in all its aspects. To do so it brings together a large number of scholars from many different disciplines. One motivating force for these scholars has been an awareness of the inadequacy of their particular scholarly discipline to provide satisfactory answers to the complex problems which they were handling both in research and in teaching. Many aspects of contemporary European economy, society and politics are indeed hard to illuminate if the light comes from only one of the traditional disciplines of academic study and this has meant that teachers, students and everybody else are frequently without adequate information on topics of immediate and important interest. The Association has therefore commissioned scholars currently working in such areas to present in short form studies of problems which are of special importance or specially noteworthy because of the lack of easily accessible information about them in current public and academic discussion. The studies are written by experts in each particular topic. They are not, however, merely for teachers and students, but for anyone who may wish to find out something further about subjects which are now much discussed but about which real information is still hard to come by. In this way the Association hopes it may bring closer what are often the separate worlds of academic and public knowledge while at the same time providing a service to readers and students in a relatively new field of study.

Alan S. Milward
University of Manchester Institute
of Science and Technology

I The Political and Economic Origins of the Common Agricultural Policy

Introduction

From the outset it was clear that the authors of the Treaty of Rome had in mind much more than an economic alliance amongst separate member countries. The preamble to the Treaty states first that the heads of state are determined to establish the foundations of an ever-closer union among the European peoples. This closer union has political, diplomatic and cultural aspects as well as those which relate purely to economics. However, the chosen instrument of integration was economic union.

Economic union was to be pursued through the market. In essence trade and the movement of resources were to be free throughout the Community and enjoy a common level of protection against goods originating elsewhere. Agriculture was ill-adapted to this approach. Protection had been given not only by customs duties but also by a variety of agricultural policies. It thus became inevitable that the Community should concern itself with agricultural policies within member countries. Such policies, which were in effect non-tariff barriers to trade and shielded internal markets from volatile world prices, were inconsistent with free trade within the Community.

There were many forms of intervention among the countries which made up the European Community (EC). Governments generally sought to restrict imports and promote exports of agricultural goods; some gave additional sums to domestic agricultural producers to supplement receipts from the market; some sought to reduce factor costs; some permitted what were in effect monopoly organisations to operate on behalf of farmers; many exempted agriculture from certain taxes and reduced the incidence of other forms of taxation upon the industry. Very commonly they provided both technical and managerial advice to farmers and invested substantial public funds in agricultural research. In the interest of consumers, governments monitored the safety of food. As a result of all measures of this type the location of agricultural production and the flow of agricultural trade depended not simply upon competitiveness but upon the action of governments. Had the Community simply allowed agricultural trade to take place between member countries without controlling the activities of member governments, the agricultural industry of the EC would have assumed a pattern determined not by its efficiency but by the willingness of the public, in differing situations, to finance agricultural support.

To avoid such distortions within a united Europe agricultural policies should have been dismantled entirely. Had this occurred, as was planned for other sectors, competitive forces would, in time, have produced a new distribution of agricultural production. This would have reflected the opportunity costs of agricultural resources in various regions and the productivity of farming in each area. Such a policy would have captured the economic benefits of creating a common market. To understand why the Community did not choose such an apprcach for agriculture we must examine:

(1)   the reasons for agricultural policy;
(2)   the differing situations of agriculture within the member countries;
(3)   what it was hoped a common policy for agriculture could achieve within the Community.

(1)   The reasons for agricultural policy

Virtually every developed country has an agricultural policy. The rationale for intervention differs from country to country and within countries from time to time but, in general terms, can be classified under one of three headings: concern for food supplies, economic benefits from agricultural policy and the welfare of the rural population.

Concern about food supplies involves both quantity and quality. Fortunately among developed countries food is seldom scarce, although there is often a threat that it may be so in war.

Failures in harvests, particularly if these occur simultaneously in several important producing areas, may lead to shortages of particular food items. The most critical occur when cereals are scarce. Cereals form the basis of many foods consumed directly by humans and of most animal feedingstuffs. Thus scarcity leads to high prices not only for bread and cakes but also for meat, eggs and milk. Governments therefore seek both a nutritionally adequate supply of food and the avoidance of sudden shortages which cause food prices to rise. In so far as food supplies from abroad may be regarded as more vulnerable to political or military interference, food security may be sought by producing more at home.

Most governments claim that the agricultural policies they pursue are of general economic benefit. Several agruments are advanced. One of the most attractive is that the actions of the government improve the efficiency with which the resources are used within the economy. Agricultural policies which provide technical and managerial advice and finance research may come into this category. More subtly it may be argued that, since many industries enjoy a measure of protection provided through their relatively monopolistic structure, agricultural support gives the farmers a degree of countervailing power. This ensures that they receive a more appropriate share of national resources.

Another group of economic benefits claimed for agricultural policy relates to trade. In countries which are substantial importers of agricultural products higher domestic production is seen as a means of import replacement. In those countries which are more self-sufficient, agricultural policy may be favoured as a means of promoting exports. This can, it is argued, enable a higher rate of economic growth to be maintained in the economy as a whole. The benefits accrue not only to farmers but also to the entire community.

Much food reaches consumers in a processed, packaged, preserved form. This has added to the diversity of human diet and probably improved its quality. It also minimises the effects of shortages of particular food items. However, since the layman (or housewife!) may be unaware of the techniques employed in adding 'convenience' to foods, it raises questions concerning the safety of additives, the reliability of some processes and the safe life of many preserved products. Adequate official monitoring is in the interest of the food industry as well as the consumer. A single case of illness as a result of eating one branded food item will have far-reaching effects on demand. However, such regulations can also restrict trade if they differ between countries.[1]

Agricultural protection is sometimes claimed to contribute to the maintenance of full employment. It is argued that many of the resources committed to agriculture would be unemployed or even unemployable elsewhere. In the absence of policy such resources might have a zero output. Thus the true 'opportunity' cost of supporting agriculture is much lower than calculations which value farm resources at current market prices suggest.

Probably the most important rationale of agricultural policy in developed countries is that it maintains the welfare of the rural population. In many rural areas agriculture is the principal economic activity. Because of farming it it profitable to provide transport and local shopping facilities. Because the rural population is reasonably large, services such as education and health can be made available at relatively lower per capita costs. If farming declines, the whole economic life of rural areas must contract, leading ultimately to outward migration and depopulation. Thus, by maintaining farm incomes, agricultural policy can sustain the life of whole rural communities.

In recent years in Europe there has been a persistent tendency towards a relative deterioration in farm income. In countries where incomes are already reasonably high, additional per capita income does not lead to a corresponding increase in the quantity of food consumed. Consumers may upgrade their diet by introducing more meat, more variety or more convenient foods. However, total demand for farm products grows less rapidly than incomes as a whole. In the EC the income elasticity of demand for food is less than unity.[2] This means that the revenues of agriculture grow less rapidly than those of other industries. If costs were to remain unchanged this would lead to a widening gap between farm and other incomes. But costs do not remain the same. Other industries, whose revenues are growing more rapidly, compete for the resources which farmers use. Sometimes this is direct competition for manpower or for land. More often it is competition for resources used to produce inputs which farmers buy: tractors, combines, fertilisers, pharmaceuticals, and so on. These costs tend to rise at a rate proportional to the growth in national income. Since this is more rapid than the rate at which the revenues of agriculture increase the implication is that, in the absence of increased productivity or government aid, farm incomes may actually decline.

Individual farmers respond by improving their farming methods. As farm management improves the productivity of land and of manpower rises. However, the demand for most agricultural goods is price inelastic.[3] Thus, as more goods reach the market prices tend to fall. The aggregate effect of individual farmers safeguarding their own positions by increasing output is to bring further downward pressure to bear on prices. Those who can but have not yet applied the improved methods will then be under severe pressure to do so. Since this will again increase output prices will fall further. Farmers who cannot apply the new methods, either because their farms are too small or because they have too little capital or because they do not possess the necessary skills, will find it impossible to compete. Either they must accept a reduced standard of living or they must leave the industry.

In purely arithmetical terms a satisfactory rate of outward migration would restore the level of incomes of those who remain to parity with other groups. Although very many people have left agriculture in Europe during the past twenty years (see Table 1), incomes still lag. Reasons why more have not left are not hard to discover. Many farmers are remote from alternative employment. Their present livelihood ensures not only a money income but also a house, some perquisites in kind and a place within a community in which their occupation gives them status. Further, although their incomes may compare unfavourably with those of workers in other sectors, economic growth, especially if accompanied by inflation, tends to raise the price of agricultural land. Thus, those who own their land may find that they are relatively rich in terms of capital although poor in terms of income. Immobility is also a feature of the relatively elderly who find it harder to re-train for new occupations and for whom employers are likely to be reluctant to provide training. Many owners of farms are relatively old. Adjustment to urban life is difficult for those accustomed to agriculture. Faced by a community with which they may have few personal contacts, an industrial discipline quite different from that they have experienced and relatively high prices for housing and food, many who migrate face, at least in the short term, periods of discomfort.

Such welfare problems have affected many citizens within the EC. When the Community came into existence, almost one person in five worked in agriculture. Economic growth was expected to continue at relatively rapid rates. To abandon all national agricultural policies and leave the industry to the full rigours of a competitive market was politically and socially unacceptable.

(2) Agriculture in member countries of the Community

Tables 1 and 2 show that agriculture varies in importance within the economies of the member states. The proportion of population engaged in farming has been falling in all member countries. During the period of rapid economic growth in the 1960s and early 1970s, many people leaving farming found jobs elsewhere. Indeed, this transfer of labour contributed to economic growth. Agricultural output continued to rise but the social implications in rural areas – ageing populations and decaying villages – were less welcome. In the 1970s, although growth has slackened, the social dimension has assumed added weight. Competition within agriculture may now displace people who have no prospect of moving to another job.

National variations in the share of population engaged in agriculture and in the proportion of Gross Domestic Product derived from it form one of the most intractable problems of the Common Agricultural Policy (CAP). Such differences imply that transfers to farmers from non-farmers financed by the EC must in aggregate benefit some member countries and represent a net cost to others. Equally the urgency of the 'problems' of agriculture is perceived very differently in the several member countries. Thus the seeds of conflict are sown.

Member countries also differ in terms of the proportion of total agricultural needs produced at home. Table 3 shows self-sufficiency ratios for various products for member countries. It is clear that a major concern of Ireland, France and the Netherlands must be markets for their exports, whilst the United Kingdom, Germany and Italy have a greater interest in keeping down import prices. For Germany and Italy, however, the political weight of farm interest groups emphasises the importance of sustaining home production, and may justify higher import prices. In contrast in the UK the relative weight of the consumer interest has tended to keep prices down, even if this means a smaller British agriculture.

The structure of agriculture within member countries also varies. Table 4 shows that most farms are small. Measures of farm area provide only a rough guide. Land quality varies so, for example, the productive potential of a large hill farm may be much smaller than of a modestly sized farm in a favoured lowland area. In some member countries farms are usually in one or two blocks of land. In others many are badly fragmented. The productive potential of a farm depends to an important degree on the capital at its disposal. Measurement of available capital even within a single farm business is extremely difficult. Comparisons between countries, where price levels and farming practices vary considerably, are all the more elusive. However, it is clear from such information as that given in Table 5 that such relatively simple criteria as the number of machines employed show considerable differences between member countries. Apart from the physical and financial aspects of farm structure the performance of agriculture may be affected by the system of land tenure. In some parts of the Community the characteristic pattern has been peasant proprietorship. In others relatively large farms have been held on a landlord/tenant basis. In yet others systems of share-cropping[4] have been common.

Within the Community the proportion of livestock production to crop production is much greater in the north than the south (Table 6). Further, there are considerable variations in the types of livestock and crops produced. Implicitly this means that changes in prices of particular products have differential effects on the fortunes of farmers in the several member countries. Prices of products vary among the Community countries.[5] The figures given in Table 7, based on figures compiled by FAO ECE, give some guide to the variety of experience among member countries since 1958. They are not a precise measurement of prices received by any particular farmer. Nevertheless, it is interesting to notice that despite the CAP the gap between countries in the earliest year and the latest year has not noticeably narrowed.

(3) What the Common Agricultural Policy hoped to achieve

Articles 38 to 45 of the Treaty of Rome concerned agriculture.

They were designed to ensure that a common market could extend to agriculture and trade in agricultural products. Article 39 sets out the objectives of the Common Agricultural Policy. These are:

(a) to increase agricultural productivity by developing technical progress and by ensuring the rational development of agricultural production and the optimum utilisation of the factors of production, particularly labour;
(b) to ensure, thereby, a fair standard of living for the agricultural population, particularly by the increasing of the individual earnings of persons engaged in agriculture;
(c) to stabilise markets;
(d) to guarantee regular supplies;
(e) to ensure reasonable prices in supplies to consumers.

In working out the Common Agricultural Policy and the special methods which it may involve due account is to be taken of:[6]

(a) the particular character of agricultural activities arising from the social structure of agriculture and from structural and natural disparities between the various agricultural regions;
(b) the need to make appropriate adjustments gradually;
(c) the fact that in member states agriculture constitutes a sector which is closely related to the economy as a whole.

The Treaty allows for common organisations of agricultural markets which may take the form either of common rules concerning competition, compulsory co-ordination of the various national market organisations or a European market organisation.

The Community was required to hold a conference shortly after the signing of the Treaty in order to provide a basis upon which the European Commission could make proposals for a Common Agricultural Policy. Agreements were ultimately to be reached by a qualified majority but unanimity was still required at the outset.

The formal language of the Treaty carefully avoids the sort of specific guarantees which farmers and others sometimes seek to read into policy documents. However, it seems fair to indicate what appeared to be the intentions of the Community at the time the Rome Treaty was signed. These included:

(a) an unimpeded flow of trade among member countries;

(b) encouragement of efficiency in agriculture which would reduce costs;

(c) an improvement in the incomes of those who work within the agricultural sector;

(d) secure supplies, stable markets and reasonable prices for consumers.

The Treaty wisely left unresolved the question of how these doubtfully compatible goals were to be achieved. Its proposals included suggestions such as long-term contracts between member countries. However, what eventually was adopted and now operates depended not on the Treaty itself but upon decisions of the Council of Ministers. In the next two chapters we look at what instruments of agricultural policy were chosen by the Council of Ministers and how they have been applied. These instruments which form the CAP as it actually exists fall into two groups: those concerned with prices and those designed to improve agricultural structures.

Notes: Chapter 1

1 Similarly differences in hygiene and veterinary regulations justified to protect the health of plants or animals may, if they differ between countries, limit trade between them.

2 $\text{Income elasticity of demand} = \dfrac{\text{Percentage change in quantity demanded}}{\text{Percentage change in income}}$

3 $\text{Price elasticity of demand} = (-1)\dfrac{\text{Percentage change in quantity demanded}}{\text{Percentage change in price}}$

4 Under this system of land tenure, annual rent payments made by the tenant to the landlord are in the form of a fixed share of the main crop output rather than a fixed sum of money. It tends to restrict the tenant-farmer to production of a particular crop unless new agreements can be negotiated with the landlord. However, in bad seasons the risks of poor crop yield or failure are shared. The main disadvantage of the system is that it diminishes the inducement for the tenant to increase production since the landlord shares in the benefits but not in the variable costs of production.

5 In practical terms there are severe difficulties in measuring prices. First, prices for agricultural products commonly vary from year to year according to changes in local market circumstances. In any particular year comparison can mislead. Secondly, prices vary according to the stage in the marketing process at which they are measured. Those measured as the goods leave the farmgate will normally be lower than those measured at wholesale or retail market. Thirdly, price measurements may refer to slightly different qualities of a product or different types of a product. Thus, beef may comprise anything from old cows to fresh prime 2-year-old steers. International comparisons are even more complicated because of the existence of different currencies and the need to convert these to a single unit. Since the value of any numeraire currency may itself be moving up and down over the period concerned this can distort changes amongst the countries compared.

6 Article 39b.

2 How the Price Policy Works

Introduction

The price policy is concerned with maintaining internal prices for EC farmers. It is based on separate price regimes for most major products and the price levels are decided annually by the Council of Ministers. In formulating the policy, the Council of Ministers has given priority to four main principles:

> the free movement of agricultural goods within the Community;
> common prices;
> a common tariff wall against Third country imports;
> joint financial responsibility.

This chapter will present a general statement of the object and methods of the price policy and the price-fixing process of the CAP, a brief review of the regulations for several major agricultural products and an indication of how the policy has developed and the problems it faces in the economic environment of the late 1960s and the 1970s.

General résumé of the system: object and methods

The price policy is the single most important mechanism of the CAP. It is based on controlling markets to achieve a desired level of price. For each product covered by the CAP, a 'common market organisation' was introduced. Over a transitional period, differing national price levels were to be brought in line so that a common price for each commodity should apply throughout the

Community. This involved a substantial realignment of farm
prices between member states. Common prices and free trade
were achieved for cereals in 1967 and for other major products in
1968. In the absence of a common European currency, from
1962 'common prices' were expressed in agricultural units of
account (UA). It was not an actual currency but a common
measure of value of agricultural products. The initial value of the
UA was equivalent to 1 United States (US) dollar or
0·888670888 grams of fine gold.[1]

The price policy has two main dimensions.

INTERNAL

A target, market price level representing a desired market price is
set for most products annually. Should the market price fall, a
'floor' price (or intervention price) for selected major products
becomes operative in the domestic market; in the event of surplus
production commodities may be bought by intervention agencies
to maintain a minimum wholesale market price level.

EXTERNAL

A common barrier operates in the form of minimum import prices
(threshold prices) set annually for goods from Third countries.
Variable import levies compensate for the difference between fixed
Community prices and fluctuating world prices. This protects the
Community market from price fluctuations and from competition
from low-priced imports. In the event of an EC surplus in a
particular commodity, the relevant management committee (see
next section) may authorise subsidies on exports of that
commodity to Third countries. This export subsidy or
'restitution' allows Community-produced goods to be competitive
in world markets even though EC domestic agricultural prices are,
in most cases, above world prices. The amount of the restitution is
determined by consideration of factors such as the level of world
prices and the amount of surplus to be shifted. It may also be
subject to political considerations and government manipulation of
the market.

Working the Policy

THE NEGOTIATION OF PRICES

The common official price levels, such as target, intervention and threshold prices, are set annually. The annual agricultural price-fixing is the occasion of a major debate on policy issues. As such it must pass through an established sequence of discussion before final approval. The structure of the Community decision-making process is summarised in Figure 1.

The Commission, as the originator, and the Council of Ministers, as the final arbiter of policy, are together the hub of the process, but no one body has ultimate power to take decisions. Any draft proposal by the European Commission is subject, after consultation with various advisory bodies or pressure from interested parties, to discussion and amendment.

Contact begins with the national governments. The Directorate-General for Agriculture invites advisory working parties drawn from experts from national governments to assist the Commission in drafting proposals. The draft is submitted for discussion to committees and farmers' representatives, e.g. COPA[2], the Economic and Social Committee, a body of representatives from trade unions, employers, professional organisations, and so on; and the Agricultural Committee of the European Parliament. At any stage, the proposal may be amended. The management committees, one for each product, are an important part of the process. Any proposal is submitted for discussion and vote by the relevant committee. The proposal, after extensive examination, will finally be submitted to the Council of Ministers for decision and implementation by the Commission and member governments.

Decisions may be of two types:

(1)   those which involve additions to or modification of existing regulations and directives;
(2)   those dealing with day-to-day decisions and administration.

When agreed, Community policies take one of several forms.

Regulation: applicable in all member states and overriding national law.

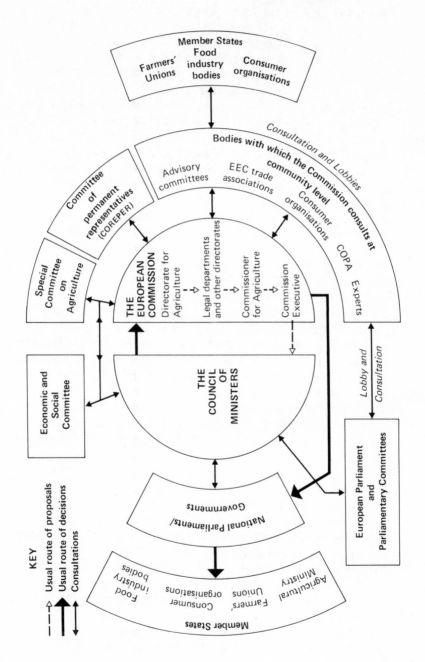

Directive:	outlines objectives binding on all member states but permits individual methods of implementation.
Decision:	is specifically addressed to a government, organisation or individual and is binding on those named.
Recommendation/ Opinions:	are not binding but generally express the Community's opinions.

The annual price review, as a regulation, is legally binding on all member states, the new prices coming into effect in all member states on specified dates. (For details of the problems faced, see later section.)

JOINT FINANCIAL RESPONSIBILITY

Responsibility for financing common measures was, from 1962, vested in the European Agricultural Guarantee and Guidance Fund (EAGGF). The Guarantee section is concerned with financing price support measures and the Guidance with structural reform measures. A programme for giving the Community independent revenue or *ressources propres* has been operative since 1971. This programme, in two phases, replaced the old regime of funds contributed solely by national exchequers. Revenue from levies on food imports and customs duties now accrues directly to the Community budget, which is in effect 'a federal budget financed by federal revenues'.[3] The principle of common financing follows on from the establishment of a common policy to ensure free internal trade and common prices.

Up to 1971 the Fund operated by reimbursing expenditure already incurred by member governments. Now estimates of future expenditure are submitted and member governments are awarded the necessary finance in advance. Total expenditure on agriculture by the Community and member states in 1977 has been estimated at around 19,500 UA. Of this, the EC through the EAGGF financed only 36 per cent. The remainder came from national sources. This element of national financing suggests that some 'uncommon' attributes of agricultural policy remain in member states.

Table 8 gives a breakdown of the Fund's expenditure. The Guarantee section accounts for the majority of funds: some 6,662 million UA in 1977.[4] The major part of this is an estimated spending of 2,545 million UA on dairy products. The Guidance section is, at present, limited to a ceiling of 325 million UA to be spent on structural improvements in production and marketing. Most of the expenditure on structural reforms and projects still comes from national governments: some 5,472 million UA on production and marketing measures, research and advisory services, and so on, and some 5,793 million UA on social measures.

Expenditure on market support measures, the Guarantee section, is estimated to rise to almost 8,500 million UA in 1979. With receipts for 1979 estimated at only 2,173 million UA, revenue generated from the CAP's duties and import levies falls far short of expenditure on agricultural support. This problem shows no sign of diminishing in the foreseeable future and, by 1981, the Community may find it difficult to meet its expenditure.

Application of the policy to individual commodities

Common market organisations exist for 96 per cent of Community farm production. The only major products not subject to a common regime are potatoes, sheepmeat and wool, and ethyl alcohol. All other major products are covered by some form of support system. The cereals system, first to be implemented, is basic to all market organisations. It is therefore dealt with in some detail. The systems for sugar, milk and milk products, meat and oil and oilseeds are then more briefly examined.[5]

CEREALS

Each year the Council of Ministers sets the target price for the whole Community. It is the price which it is hoped producers in the area of greatest deficit (Duisburg) will receive for their grain. It is increased over the season in a series of monthly increments to encourage orderly marketing.

The internal market is supported by the intervention price which runs parallel to the target price, some 12–20 per cent below it. Grain sold into intervention, once the market price drops to intervention level, must meet certain quality and quantity standards. These are designed to reduce the quantity being sold into stores and to ensure its storability. Much of the grain going into intervention goes through 'the trade' which can ensure supply of the minimum quantities required. The actual price a farmer receives will be below the intervention price after allowance is made for transport and handling costs.

Initially, the intervention price for wheat was based on its use for milling into bread flour but more wheat is produced in the EC than can be used for milling. The separate, higher intervention price made soft wheat too expensive to use as animal feed which led the Community into the financially and politically undesirable process of 'denaturing' wheat. In effect, this amounted to adding a dye so ensuring that milling wheat would be unsuitable for bread-making. It could then be sold at a lower price for animal feed. The difference between the intervention price for milling wheat and its value as feed was paid as a 'denaturing subsidy' by the EC. The intervention price for wheat has been eliminated and the market is now supported by a 'reference' price set at a level 13 per cent above a common intervention price for feed grains (barley, maize and feed wheat). Home-produced milling wheat remains too expensive to use as animal feed and, depending on the quality, is often unsuitable for bread-making for human consumption. The Community, therefore, has had to import large quantities of wheat; some 4 million tonnes in 1977/8. At the same time, it is paying for storage of just under 7 million tonnes of surplus home-produced wheat and subsidising export of around 4 million tonnes.

A system of variable import levies protects the Community market from cheap grain from Third countries. A threshold price, linked to the target price at Duisburg, is determined annually for all entry ports to the EC.[6] A levy on imports is imposed when the world price is below the EC threshold price. When world prices are higher than the EC prices, an equivalent tax may be placed on exports. The import control is effective in maintaining a stable internal market price provided grain is being imported. When internal production exceeds internal demand either intervention is needed to support the price level or restitutions may be offered on exports of the surpluses on to world markets since Community prices are usually higher than world prices.

MILK AND MILK PRODUCTS

This market has certain unique features. There is an overwhelming problem of oversupply at established prices. Only about one-quarter of the milk produced in the Community goes for liquid consumption; emphasis is therefore on manufactured milk products. The target price set for milk delivered to the dairy is supported by intervention in the butter, skim milk powder and cheese markets. No intervention price exists for liquid milk. Dairies pay producers a price which reflects what they receive for all their sales. Consequently the prices paid to producers have often been up to 15 per cent below the target price. Protection from low-priced imports is ensured by threshold prices established for twelve 'pilot' products. Threshold prices for other products are derived from these.

Continued overproduction has created large surpluses of milk products, especially skim milk powder, within the Community. The size of such stocks varies from year to year but the underlying problem has troubled the Community since its inception. Several attempts have been made to reduce the problem. As early as 1968, the Mansholt Plan (see Chapter 3) argued that resources should be redirected from milk production. Attempts have also been made to encourage dairy farmers to switch to beef. In 1977 a 'co-responsibility levy' was imposed on producers, in effect reducing prices to farmers whilst leaving market prices unaffected. These policies have to date proved ineffective, not least because of the many small farmers who derive their income from milk production and for whom no alternative work or source of income exists.

MEAT

Beef and Veal
The beef and veal regime also attempts to maintain Community market prices as close as possible to a price level decided upon by the Council of Ministers. This 'guide' price is the focal point of the regime. It is set annually, taking account of the situation in the market for milk and dairy products as well as the outlook for beef and veal production and consumption. Intervention operates when the average wholesale price (known as a reference price) falls below 95 per cent of the guide price. The intervention price itself

is 90 per cent of the guide price. The official prices, supposed to take effect from April, now usually form part of the annual price review and apply much later in the marketing year.

In addition to measures to stabilise the internal market through intervention and variable import levies, the Community provides protection for domestic producers of beef through customs duties (the 'ad valorem' tariff). These amount to 16 per cent on imports of live cattle and calves and 20 per cent on frozen and chilled beef and veal imports.

The beef and veal market has marked cyclical tendencies which extend over several years: for example, shortages and high prices in 1972/3 were followed in 1974 by excess supplies and low prices. This volatility in the market increases the problems of price-fixing as prices cannot be anticipated over the medium term with any certainty. In 1974, in an attempt to introduce stability and as a result of the UK reluctance to support the depressed market, a system of direct seasonal payments, known as the 'variable premium' for beef cattle, was introduced to operate in conjunction with intervention buying. Although this was a Community scheme it operated only in the UK.

The cyclical problems of regularising the beef market are further complicated by interactions with other products. As most EC beef is produced from the dairy herd, changes in the policy for milk may influence the production of beef. Other meats, for example, pork, compete in the market place. Costs of pig production are, in turn, affected by alterations in cereal prices and hence the competitiveness of pork in the meat market is affected.

Pigs and Poultry

The more volatile nature of this market necessitates a more flexible approach than that for beef and veal. There is no guide price; instead a 'basic' price is set. If the reported market price, the reference price, falls below this level, intervention may be undertaken. Intentionally the intervention price, set at 85-92 per cent of the basic price, has been made unattractive to producers. Up to 1974 it only operated once. A 'sluicegate' price, which represents the calculated costs of producing pigs abroad, regulates imports of pigs and pigmeat which are offered above the Community's threshold price.

There are important differences in the approach to regulating the market in pigs compared to the regimes for cereals, milk or beef. Prices are allowed to move more freely, the industry being expected to take corrective action should prices fall or rise unduly. The logic behind this difference is partly technical: pig numbers can be expanded relatively rapidly. In part it is economic: pigs are seen as 'processed cereals' so that controlling the cereal market should itself involve a degree of stability in pig production. It is also political: pig production may form a small part of the activity of many farms but the relatively large modern 'factory' units have little political influence compared with cereal-growers or milk-producers.

Similar arguments apply to an even greater extent to intensive poultry systems. The Community does not operate any overall support mechanism. It ensures Community preference by variable levies, a sluicegate price and customs duties but relies on the industry to respond to internal market pressures.

SUGAR

The sugar market is subject to far more direct intervention and stricter control than other markets. The reasons for this lie largely in the fact that the Community has, under pressure from sugar producers, fixed a price well in excess of that needed to supply its markets. Community production exceeds its needs: in 1977–8 some 3·3 million tonnes were exported (most with a common subsidy). This has been accentuated since 1973 when it agreed, under the Lomé Convention, to import some 1·3 million tonnes of cane sugar annually from developing countries. The surplus of sugar is such that the cost to the Community is very high.

The sugar regime is based on quotas, support buying and penalties for overproduction. A target price is set annually. Intervention at the full amount supports only a limited quantity of sugar, the 'basic' or 'A' quota, 9·136 million tonnes for the whole Community. Above this basic quantity, a further 'B' quota has been fixed at 35 per cent of the A quota. Sugar produced above the A quota but within the higher B quota is supported only at 70 per cent of the full intervention price. The differential between the two quotas is intended to discourage overproduction. Sugar produced beyond the B quota receives no official support.

Apart from imports allowed under the Lomé agreement, sugar from Third countries is subject to variable levies based on threshold prices. The effect has been to virtually block such supplies. The EC is not a member of the International Sugar Agreement and so has no obligation to accept any further sugar imports.

OIL AND OIL SEEDS

Olive oil is the only oil in which the EC is substantially self-sufficient. Currently, 70–80 per cent of the EC needs are met by home production. A common organisation exists for oils because of the commercial interrelationships between animal fats, butter and edible oils. Rape, sunflower seeds and soyabeans are the main oilseeds covered by common regulations. There is a wish to stimulate their production to cut down imports. The characteristics of the regime differ from those considered so far because the EC is a large net importer.

Imports of some oils are subject to customs duties but there is no other control. Olive oil imports are subject to variable levies, that is, the difference between the threshold price and a c.i.f. import price. Intervention operates to support a derived market target price for the Community. As internal supplies are subject to fierce competition from imports, a deficiency payment is made to producers to maintain a producer target price. The producer target price is intended to ensure an adequate volume of production; the market target price is set at a competitive level with other oils. The difference between the higher producer price and the market price is offset by the subsidy paid to producers.

For oilseeds, seasonally phased target (i.e. wholesale) and intervention prices are set. Different intervention prices are derived for various regions of the Community. As a result of the agreements under the General Agreement on Tariffs and Trade (GATT) imported oilseeds have free access to Community markets. A deficiency payment or subsidy representing the difference between a higher Community target price and the lowest c.i.f. price is payable to producers, wholesalers or oilseed-crushers in order to give producers a return closer to the target price. The subsidy is normally claimed by oilseed-crushers who, theoretically, pass it to producers in the prices they pay for oilseed. This is to ensure an adequate supply of oilseeds by guaranteeing an adequate producer return.

The Development of the Policy

During the 1960s, a period of economic growth, when the pricing system was devised, it was possible to envisage its successful operation. Until 1969 each currency had fixed exchange rates against the US dollar. Since the differing national currencies had fixed exchange rates, changes in UA prices would apply equally in all countries. The rate of exchange between the UA and national currencies became known as the Green rate or 'representative' rate and formed the basis of what was termed 'Green money'. In the late 1960s, pressures began to develop as the economic progress of member states, notably France and Germany, diverged. In 1969, France devalued the franc, but due to the effect on consumer prices, refused to devalue the Green rate. Germany subsequently revalued but was equally unwilling to decrease prices to its farmers. In the absence of further measures trade distortion would have resulted, as speculators, buying in Deutschmarks, sold cheap French goods in Germany, so achieving more favourable rates of exchange for the Deutschmark. This was unacceptable to the Community. As a purely temporary measure the rate of exchange used for agricultural purposes, the Green rate, and the market rate of exchange were allowed to diverge. To compensate for the discrepancies in currencies, a system of Monetary Compensatory Amounts (MCAs)[7] was introduced. They operated as levies on French (i.e. devalued currency) exports and subsidies on its imports. The system applied in reverse for Germany.

The use of MCAs expanded and continued into the 1970s in an effort to maintain unity of markets after all European currencies 'floated' in 1971 and currencies were no longer tied by a fixed exchange rate to the US dollar. The UA itself remained tied to the US dollar until 1973 when the enlargement of the Community posed further problems. The UA was then revalued using an average of several European currencies, those of Germany, Denmark, Belgium, Luxembourg, the Netherlands and, originally, France – the 'Joint Float' currencies.

MCAs have allowed Community trade to continue in the absence of common pricing but they have permitted the 'real' level of prices (i.e. how many other goods a given quantity of farm products will buy) to vary among member countries. They have allowed growing distortion of production as farmers in countries with strong currencies (e.g. Germany, Denmark) are paid more than farmers in countries with weak currencies (e.g. France, the UK). Further, MCAs have represented an ever-increasing share of the EEC budget. Nevertheless, some important elements of unity remain. Changes in UA prices apply to all countries and the Community can still give preference to its own producers *vis-à-vis* Third country supplies.

As a result of the economic pressures and monetary disorder, the Community introduced in 1979 the European Monetary System (EMS). At the centre of EMS is the European Currency Unit (ECU). The old agricultural units of account are equal to 1·21 ECU. The value of the ECU is based on a currency 'basket'; this links all the currencies of the Nine on a basis which reflects differing shares in EC output. It is hoped that this common action, if it receives adequate support, will lead to greater monetary stability and remove the domination of the unit of account by the stronger currencies.

Price-Fixing

The annual price-fixing, when official price levels for agricultural products are set, has posed major problems for the Community. Negotiations, often protracted, have placed strains on the Community's decision-making process and exposed the difficulties of maintaining unity. The difficulties can be attributed to several major issues.

(1) THE APPROPRIATE LEVEL OF FARM PRICES

Theoretically, in a free market, the appropriate level of agricultural prices is that which equates supply and demand, that is, clears the market. Agricultural production is unpredictable and subject to large fluctuations leading in turn to fluctuations in price. Attempts by governments to regularise prices and so farmers' income levels do not remove this difficulty. If pre-set prices are too high, and administratively prevented from falling, costs fall on taxpayers or consumers. If they are too low, farmers' revenues will be affected and production of some products may decline to a level at which shortages occur. Since governments do not have perfect knowledge about future production or markets they tend

to err on the side of caution, fixing prices which lead to over-
rather than underproduction.

In addition to these difficulties, the EC faces other problems
which affect price determination. First, the price fixed has a social
function. If the smaller farmers receive enough to survive, the
larger often receive too much and surplus production is
encouraged. Secondly, the policy implies transfer of income not
only between consumers, taxpayers and farmers but also between
member states. This results from differences in the character and
strengths of national economies as well as differences in farming.
Those who gain from the policy are anxious to resist change, those
who lose demand reform. Thirdly, any decision on price levels
requires virtual unanimity in the Council of Ministers. The
political strength of the farm lobby varies in member countries.
Any marked change in prices is bound to be politically
unacceptable to some country. Such deep-rooted difficulties in
identifying and applying a price appropriate for the Community
inevitably leads to prolonged and often stormy discussions within
the Council of Ministers. In effect it raises queries about the
plausibility of operating the CAP along the lines of the price policy
sketched in this chapter.

(2) MONETARY INSTABILITY

Whilst the MCA system has enabled some semblance of free trade
to be maintained in the Common Market, despite the considerable
economic divergence between states, it has also added to the
problems of price fixing. This is largely due to differing internal
circumstances of member countries and the intercountry transfers
that may result from changes in the MCA system. Germany, with
the largest positive MCA, has consistently refused to revalue the
Green Deutschmark unless an equivalent percentage increase in
farm prices could be secured. Conversely, until 1979, the UK,
with the largest negative MCA, has been reluctant to devalue the
Green pound by significant amounts, particularly since other
countries were often pressing for price increases.

(3) THE NORTH-SOUTH PROBLEM

The difficulty of maintaining unity in the face of economic
divergence was outlined in the last section. The north–south

problem highlights these divergencies and also difficulties caused by differing production conditions in each member state. A major part of the Guarantee section of the budget is spent on commodities produced in the north sector of the Community, for example, beef, cereals and milk products. Products from the Mediterranean zone receive relatively little support. The southern region contains the largest proportion of poor farmers. The price-fixing can therefore be prolonged by Italy's attempts to improve the position of these farmers whilst countries like Germany, the Netherlands and the UK seek to defend their positions.

This third issue may assume even greater importance in the face of three new considerations:

the entry of three new Mediterranean countries, i.e. Greece, Spain and Portugal;

attempts to switch more of the budget from price support to structural and regional schemes;

the probable failure of Community income to cover budgetary expenditure by 1981.

The Price Policy: an evaluation

The effectiveness of the price policy can be evaluated in terms of its success in achieving the Treaty of Rome objectives (outlined in Chapter 1). This will be examined under three headings:

(1) unification of markets and common prices;
(2) price stabilisation and income support;
(3) allocation of resources according to criteria of economic efficiency.

(1) UNIFICATION OF MARKETS AND COMMON PRICES

Common prices and free trade were achieved in the late 1960s. This adjustment involved realignment of farm prices between member states: agricultural prices in Germany, especially for cereals, declined whilst those in France and the Netherlands showed substantial increases. Cereal prices were fixed at high levels which in turn affected livestock prices and returns. This unified system of markets was only short-lived. From 1969, economic divergencies between member states threatened unity and led to the introduction of MCAs. Whilst the MCA system has enabled an appearance of unity to be maintained, it has added to the problems of price-fixing. The prolonged annual price negotiations are not only time-absorbing but they prevent realistic discussion of the longer-run reforms of the CAP.

Only by a greater degree of economic convergence among member states could the problems be solved without radical changes in the policy. The most recent round of energy price increases and the very real risk of economic recession means that the policy is likely to be under severe pressure for the foreseeable future.

(2) PRICE STABILISATION AND INCOME SUPPORT

In the 1960s the price mechanism was the sole means of both stabilising prices and supporting farm incomes. Given the rapid inflation of the 1970s, it has proved impossible to support the incomes of the smallest and most inefficient at adequate levels without raising prices to such a level as to encourage the larger and more efficient farmers to produce supluses. These structural surpluses have now reached such proportions as to impose severe strains upon the Community budget. The problem will continue as long as price support measures remain the major policy instrument for maintenance of farm incomes. Recently there has been recognition of a need for adequate regional and structural reforms (see next chapter) which it is hoped will equalise the positions of member states and reduce the budgetary cost of the policy by reducing price support.

(3) ALLOCATION OF RESOURCES ACCORDING TO CRITERIA OF ECONOMIC EFFICIENCY

The CAP has proved costly to the Community budget and most of these costs have occurred because of the price support policy. However, its cost to the Community's economy may be more serious. Because agricultural prices are higher than market conditions warrant, more resources are allocated than would occur if less protection was afforded. The loss to the economy represents the difference between the value of the output of these 'additional' resources in and outside agriculture. Calculation is not possible in the absence of knowledge about what farm workers could do in other sectors or what capital injected into farming might earn elsewhere. Such information does not exist but in a growing economy it seems probable that the opportunity cost of an agricultural policy which holds more resources in their current uses tends to rise.

Concern with budget costs is understandable. More than three-quarters of the EC budget is spent on maintaining prices. In fact this represents only that part of the financing of the CAP for which consumers do not pay. In the framework of EC aggregates it is quite modest, 0·5 per cent of the EC gross domestic product, 3·1 per cent of expenditure on food and 11·2 per cent of final agricultural production. To concentrate on this element is to miss the true cost of the price policy to the economy of the Community.

Notes: Chapter 2

1   The units of account used in this report are:

UA or AUA	The agricultural unit of account.
	The rate of exchange between the UA and each national currency, known as the Green rate, was fixed until 1973. The Green rates of exchange are now theoretically fixed annually.
UA	This also denotes the unit of account used in the Community budget until 1977. The budgetary rate of exchange between the UA and each national currency was fixed.
EUA	The European Unit of Account.
	This replaced the UA in the Community budget from January 1978.
ECU	The European Currency Unit. This replaced all other units of account from March 1979.

Details are given later in Chapter 2.

Values of the Units of Account in £ Sterling:

	£1 = AUA/UA*	£1 = UA	£1 = ECU
1973	2·1644	2·4	—
1974	2·0053	2·4	—
1975	1·86369 and 1·96178	2·4	—
1976	1·75560	2·4	—
1977	1·70463	2·4	—
1978	1·61940 and 1·57678	2·4	—
1979	—	—	1·81094 and 1·72039

*   Where two values are shown for one year the Green rate was altered during the course of the year.

2   Comité des organisations professionelles agricoles des pays de la Communauté économique européenne. This is the federation of all farmers' unions and the representative of the farming lobby.

3   European Commission, *The Common Agricultural Policy* (1977), p. 16.

4   European Commission, *The Agricultural Situation in the Community 1978 Report.*

5   For a more detailed outline of the price systems, see M. Butterwick and E. Neville-Rolfe, *Agricultural Marketing and the EEC* (1972), and EC Information Services, *The Common Agricultural Policy* (1979).

6   There are no target prices for feed grains such as oats, millet, sorghum, etc. The threshold price for these is linked to the price set for barley.

7   For more technical details, see C. Mackel, *The Development, Role and Effects of Green Money* (1977), North of Scotland College of Agriculture Bulletin No. 13.

3

The Community and the Structural Improvement of Agriculture

Introduction

Attempts by governments to influence directly farm employment, farm size and the distribution and quantity of capital are generally called 'structural policy'. This chapter covers the development of the Community's structural policy for agricultural production and marketing. The policy will be considered in two stages: 1964–7 and 1968–78, the first a period of fragmentation and development of nationalist policies, the second one of more cohesive development.

The Problem Faced in 1960

In the early 1960s there were many small marginally viable farms throughout the Community. About half the 6 million EC farms were under 5 hectares, although many of these were not full-time farms. In 1960 average farm size was just under 11 hectares (see Table 4). Although during the 1960s the agricultural working population decreased by about one-third (some 4 million people) and productivity increased by almost 7 per cent annually, this did not lead to a basic streamlining of the structure of agricultural production in terms of farm size.

For the Community, the differing circumstances of member countries made it difficult to devise a common aim for structural policies. Initially, emphasis was given to supporting those national policies which were consistent with Community objectives.[1] However, the incompatibility of some national policies with Community interests and the growing recognition that price policy alone could not resolve the difficulties of the CAP eventually led to a more positive approach.

Structural Policy, 1964–7

The first real step towards a common structural policy had been taken in 1962 when, to supplement and encourage the national governments' measures, the Community provided for a Guidance section in the agricultural fund (EAGGF). From 1964, resources from the Guidance Fund were used to finance:

(1) individual projects submitted by each member state;
(2) solutions to specific problems in an individual state or on a Community level in production or marketing (e.g. premiums for slaughtering cows and withdrawing milk products from the market; aid to Italy following flood damage).

Individual projects were defined as 'any public, semi-public or private project designed exclusively or in part for the improvement of agricultural structures'[2] and could be funded, at least in part, by EAGGF. Any project had to further the basic objectives of the CAP to:

(1) improve the structure of agricultural production, or
(2) improve the marketing structure of products subject to a common market regime, or
(3) assist in adjusting production to meet market demands.

Whilst still fostering individual action, therefore, it was a step towards common measures and co-operation throughout the Community. Aid was intended to assist those areas where the need was greatest. In reality the amounts given have proved to be closely related to budgetary contributions of each state to the common fund. The largest contributors received the largest amount of aid: Germany received most, followed by Italy and France.

From 1964 to 1966 the Guidance section of the EAGGF had available up to one-third of the expenditure of the Guarantee section. From the outset structural policy was always a smaller element in the CAP than price and marketing support. By 1966 total expenditure on price policy exceeded expectations and from 1967 a ceiling was put on the Guidance section of 285 million UA (= £118·7 million). This was raised on the accession of the new

member states in 1973 to a maximum of 325 million UA
(= £135·4 million).[3]

From 1964 to 1970, 292 million UA were allocated to
individual projects for improving structures covering such
schemes as land allocation, improvement of rural infrastructure,
irrigation, drainage, and so on. A further 192 million UA were
allocated for improvements in marketing structure. Joint
production/marketing projects received 25 million UA.
Financing of individual projects was regarded as a temporary
measure but, despite the wish for a more cohesive policy, this
loosely controlled development of national policies continued into
the 1970s.

The Mansholt Plan: 1968 and beyond

By 1967, agricultural structure in the Community showed little
improvement. The average size of farms of over 1 hectare had
taken three years to increase by 1 hectare in France, and in
Germany seven years to increase by 0·5 hectares. In total,
throughout the Six, there were in 1967 only around 170,000
farms with more than 50 hectares of land.[4]

The realisation that the situation was made worse rather than
improved by a price policy constrained by the need to maintain the
incomes of existing producers prompted a fresh look at structural
problems. Already dramatic evidence of the inadequacy of price
policy was provided by the accumulation of surplus milk products.
The Commission examined existing arrangements and the
resulting *Memorandum on the Reform of Agriculture in the EEC*
(commonly known as the Mansholt Plan) heralded a new and
more comprehensive attitude towards tackling agricultural
structural problems.

The report noted that among the problems facing agriculture
were the small size and fragmentation of farming units. In
particular milk production was concentrated on such small units,
making it politically unacceptable to correct oversupply by cutting
prices. Finally, the relative bias in the agricultural population
towards the upper age-group retarded any adjustment process.
The dimensions of the problem were daunting. Structural
improvement could not take place until a large proportion of the
12 million people still working in agriculture left the industry.

Unless this happened, agriculture would remain a 'social problem'. Small farmers, maintained by the price support policy, would continue to produce commodities for which the market demand was saturated. Low incomes gave them no alternative but to try to increase their own output even if the market was oversupplied.

The Memorandum outlined a comprehensive set of proposals for the reform of agriculture. It proposed reform of the structure of production; a reduction of the total agricultural area; and marketing improvements. The 'new' structure was to be based on forming 'enterprises of adequate size' by reducing the size of the agricultural population and by increasing the size of farms. Such modernised units would, it was argued, be able to respond to market conditions.

Two kinds of measures were proposed:

(1)   to help people to take up alternative occupations or to retire;
(2)   to help people who remained to modernise their farms.

The Memorandum also sought marketing improvements, particularly in information, the organisation of producer associations and greater responsibility for farmers in the marketing of their produce. It argued that larger, modernised farms could play a more responsible role in marketing their output. This would ease the strain on price policy. These comprehensive and, in many ways, drastic proposals were designed to shock the Community into realising the inadequacy of short-term price measures in the context of a longstanding and self-perpetuating social structural problem. It provoked widespread reaction and much lively discussion but no positive action was taken until 1972 when the three directives[5] were introduced. These represented a noticeable retreat from the initial plan, though they retained many of its proposals. They were concerned with:

(1)   farm modernisation;
(2)   encouragement to specific farmers to leave farming and to allocate their land for the improvement of remaining holdings;
(3)   measures for training and advice to farmers.

THE MODERNISATION OF FARMS (DIRECTIVE 72/159/EEC)

This directive aims to improve the living standards of the agricultural population throughout the Community by a comprehensive modernisation programme for some farm holdings. Aid is available for investment in the unit for the implementation of an approved 'development plan'. The development plan, covering a period of not more than six years, must be submitted to the national government. By the end of the period, the holding should be able to employ one man full-time, providing him with an imcome comparable with average non-agricultural earnings in that region. The selectiveness of the scheme means that aid is available only to those farmers who satisfy certain criteria.[6] To the farmers who successfully comply with the conditions of the directive, several forms of aid are available:

> aids for the investment necessary to implement the development plan either as an interest rebate or as a capital grant;
> priority to acquire land released under the second directive (see below).

For eligible claims, EAGGF will normally grant up to 25 per cent of the aid, the balance to be paid by the member state. In some situations, where the structural problem is especially severe, a larger proportion of expenditure may be met from EC sources. The directive does not prohibit the granting of national aid to holdings not falling in the categories it defines. It does, however, severely restrict such aids except on a regional or temporary basis.

In the UK this directive is implemented through the Farm and Horticulture Development Scheme whose provisions coincide with those contained in the directive. This scheme will operate until 10 April 1982.

THE CESSATION OF FARMING AND THE REALLOCATION OF THE UTILISED AGRICULTURAL AREA FOR STRUCTURAL IMPROVEMENT (DIRECTIVE 72/160/EEC)

This directive aims to encourage farmers whose holdings are incapable of providing an adequate income to give up farming and release their land for reallocation. It does so by providing financial assistance.

To qualify for a retirement payment, the farmer/farm worker must have worked in agriculture for the five years prior to submitting his application and, during that time, he must have devoted at least 50 per cent of his working life to agriculture and derived at least 50 per cent of his income from it. Land released must be reallocated to increase the size of 'developing' farms (i.e. those subject to a development plan under Directive 72/159/ EEC), or withdrawn from agricultural use for afforestation or for recreational or public use. The land may be sold to its new user or rented for a minimum of twelve years. The directive provides a pension for retiring farmers which varies with the age or income of the farmer and the region in which he farms (e.g. in Ireland and most of Italy younger farmers are eligible and the EAGGF's contribution is higher than in more favoured regions).

To comply with this directive, the UK operates two schemes:

Farm Structure (Payment to Outgoers') Scheme (1973);[7]
Farm Amalgamation Scheme (1973).[8]

Other countries have their own schemes and national governments are allowed under the provisions of the directive to award additional aid at their own expense.

THE PROVISION OF SOCIOECONOMIC GUIDANCE FOR THE ACQUISITION OF OCCUPATIONAL SKILLS BY PERSONS ENGAGED IN AGRICULTURE (DIRECTIVE 72/161/EEC)

This directive aims to develop information services and professional advice to farmers, farm workers and their families. Information and guidance is intended to deal with matters relating to agriculture and with more general aspects concerning the economic and social conditions of the farming community. It is intended that this guidance will give the farming community a better understanding of its situation and opportunities; help to improve farming skills if they choose to remain in agriculture; and help individuals to adapt to new situations should they decide to leave farming.

The directive requires member states to implement the following measures:

the creation of special advisory services;
the training of 'vocational guidance' advisers;
the setting up of training centres to provide courses for farmers and hired farm workers;
the provision of an income during the training period for those wishing to re-train for an occupation outside agriculture.

The EAGGF will refund up to 25 per cent of a member state's expenditure of socioeconomic measures. Expenditure on re-training, however, will only be funded until such time as the financing of it is taken over by the European Social Fund.

This group of directives attempts to tackle three principal factors which limit the ability of the farming sector to earn an income comparable with other parts of the economy:

inflexible farm structures;
the age distribution of the farming population;
inflexibility of output which limits response to market conditions.

The directives are, however, 'blanket' measures and, as such, do not serve the needs of all regions equally. Agriculture systems are diverse. They vary both between countries and between regions within one country. Within the Community there exist substantial regional variations in farm income. Many of the areas where income is lowest suffer permanent handicaps. Natural features such as slope, poor soil type, climate and altitude of an area mean generally higher production costs and a lower return due to poorer yields. The agriculture of such regions is uncompetitive in a Community context. In a situation of surplus, competition in a 'free' market would have implied the cessation of production, depopulation and severe social and ecological consequences. To maintain prices adequate to remunerate these farmers would have led to surplus elsewhere. Such an outcome was unacceptable to the Community.

Regional differentiation is against the principles of a common market. The Community, however, in the Rome Treaty Article 39 recognised that special problems existed as a result of regional disparities. The gap between the less favoured and the more prosperous agricultural regions and anxieties about rural depopulation have forced the Commission to consider specific policy measures to assist the weaker areas. Aid to the less favoured areas has been proposed with two principal objectives:

to enable mountain and hill farmers and farmers from other poorer areas to restructure their farms in line with other more favoured regions of the Community;
to preserve the ecological balance of the natural environment for the benefit of society as a whole.

MOUNTAIN AND HILL FARMING AND FARMING IN CERTAIN LESS FAVOURED AREAS (DIRECTIVE 75/268/EEC)

This directive proposes direct compensation to farmers proportional to the extent of the natural handicaps with which they have to contend. As a result it is intended that they will continue to farm in these regions and if possible carry out farm modernisation. By supporting agriculture, it is hoped the social and ecological balance of the region will be maintained, i.e. the population will be kept at acceptable levels, the countryside will be conserved, and the continuance of farming will be ensured as the younger population will remain in the area. To be eligible farmers must work at least 3 hectares of utilised agricultural area (UAA) and continue to farm for at least five years after aid has been granted. The main measures include:

an annual compensatory allowance to the farmer connected to the severity of the natural handicaps and the volume of his business;
investment aid for farm modernisation – Directive 73/160/EEC allows for the special features of the areas to be considered, which may include croft and tourist activities;
aid for co-operative farming practices.

The directive affects some 12–14 per cent of the agricultural population of the Community who farm about 20 per cent of the total cultivated area. It has added a new regional dimension to the operation of agricultural policy.

Further structural measures: marketing improvements

After concentration on rationalising farm structure, the new policies took a further step in the form of a measure to improve the marketing of agricultural products: Regulation (EEC) No. 355/77. Aid is given to eligible projects relating to improvement of processing, preservation, packaging or storage of agricultural products. These projects should be part of approved regional, national or Community programmes. Several programmes with such projects forming a part of them have already been submitted to the Commission from four member states: Germany, the UK, Ireland and Denmark.

After many years of debate a further measure to improve marketing structures for agricultural products was adopted in June 1978 regarding the organisation of 'producer groups and associations thereof' (Regulation (EEC) No. 1360/78). This is intended to strengthen the farmers' bargaining power by fostering co-operation in the marketing of agricultural products. It is limited in application to Italy and certain areas of Belgium and France.

The Mediterranean 'package', 1978

The operation of the price policy, which favours 'northern' products, and the prospective enlargement of the Community to include Greece, Spain and Portugal caused concern about certain areas around the Mediterranean. As a result, specific measures have been adopted in what is generally termed the Mediterranean 'package'. These include:

the provision of more finance in the Mediterranean regions for promoting better marketing structures (Regulation (EEC) No. 1361/78);
the guidance of irrigation works in the Mezzogiorno (Regulation (EEC) No. 1362/78);
the improvement of public amenities in certain rural areas (Regulation (EEC) No. 1760/78);
the acceleration of restructuring of vineyards in some Mediterranean regions of France (Directive 78/627/EEC).

Further proposals have been made for the Mediterranean regions relating to forestry development, irrigation and advisory work. Similar attention has been given to certain other less-favoured regions, for example, the acceleration of drainage operations in the west of Ireland (Directive 78/628/EEC).

By 1978, the lack of success of the 1972 'compromise' directives had given rise to a movement towards a reacceptance of the much more comprehensive thinking behind the Mansholt Plan of 1968. At the end of 1977, the Commission submitted a report on the effectiveness of the 1972 structural directives together with proposals for their reform. The report argues that from 1972 to 1978 the directive on the modernisation of farms has had moderate success but the retirement directive has been an almost complete failure. The directives have been thwarted by lack of funds and by national laws which inhibit their effective operation. Whilst the measures have attempted to tighten up structure and reduce farm population, the price policy, the focus of attention and of funds, has encouraged the small, often economically non-viable, farms to 'stay in business' by giving an acceptable return through high farm prices. In the light of this analysis the proposals made seem rather modest. New proposals, if adopted, would strengthen aid to the poorest farmers, increase pensions under the retirement scheme and discriminate still more in favour of disadvantaged regions.

In following the development of structural policy in the EC, it becomes clear that the concept of agricultural policy has broadened to embrace social considerations. The main focus of the CAP has become more overtly the welfare of the farmer in relation to his counterparts in other sectors of the economy. As well as broadening to encompass social considerations, structural policy has become concerned with rectifying the divergence of regional agricultural income levels within the Community. To this end, specific measures designed to tackle the particular problems of certain areas have come to the fore. How long the regional 'slant' will continue as a major influence in policy formulation cannot be estimated.

Notes: Chapter 3

1 For a description of individual countries' structural policies, see OECD, *Structural Reform Measures in Agriculture* (1972).
2 Regulation No. 17/64, EEC *OJ* No. 34, 27 February 1964.
3 Converted at the budgetary rate 2·4 UA = £1.
4 It should be noted that farm size is only a relatively crude guide to the development of agricultural structures. Changes are nevertheless relevant.
5 The measures for structural reform are directives, i.e. the national governments of the Community are under an obligation to implement measures in order to achieve the common objectives. The actual measures used vary from country to country.
6 For details, see National Farmers' Unions, *EEC Policies for Improving Farm Structures* (1974).
7 Statutory Instruments 1973, No. 1404.
8 Statutory Instruments 1973, No. 1403.

4

The Common Agricultural Policy and the Rest of the World

Introduction

The Common Agricultural Policy regulates the terms upon which the Community trades with the rest of the world. In this chapter the nature of this relationship is discussed and attention is drawn to some of the issues which concern both the Community and other countries.

Four ways in which the Community's policy affects other countries must be distinguished.

(1) The Community, through the duties it imposes upon some agricultural imports and the variable import levies which apply to products covered by the CAP, ensures that internal producers enjoy a measure of protection from external competition. In effect the apparatus prevents price-cutting by overseas suppliers undermining the internal price level within the Community. The justification for such a policy stems in part from the problems of agricultural adjustment within the Community and in part from the residual character of international trade in agricultural products. Most industrial countries ensure that their internal price level is insulated from external prices. To do so they may provide subsidies for exports and limit competing imports. The effect is that surpluses occurring in any producing area tend to be unloaded on to the world market even if the prices received are well below the costs of production. In times of scarcity, countries resort to the world market seeking to top up domestic supplies. The result is depressed world prices interspersed by very sudden and substantial rises in price. The history of sugar prices in the 1970s provides a good

example. In 1968/9 the import price was 6·29 UA per
100 kg, by 1974/5 it had risen to 66·60 UA but by
1977/8, it had dropped back to 13·55 UA.

(2)   The Common Agricultural Policy encourages the substitution of imports by home production. A less obvious
but quite substantial form of import substitution occurs
where the level of protection accorded to different products
varies. In such a situation the less-taxed imported product
will tend to supplant the more heavily taxed imported
product in the Community's market. The most important
example of this has been the import of tapioca and soya to
feed animals in place of cereals. This not only creates
problems for Third country cereal suppliers but may also
increase the costliness of the common cereals policy to the
Community itself.

(3)   Where Community production exceeds domestic consumption the CAP provides for export restitutions. The
size of the restitutions is determined by the management
committee for the product concerned. If necessary, very
substantial export subsidies may be provided on the grounds
that this clears the Community market even though the final
revenue received from foreigners is well below the costs of
production. Such export subsidies disrupt agricultural
markets for Third countries. Thus, for example, New
Zealand may be in competition from subsidised low-priced
butter exports from the Community in Hong Kong.

(4)   In trade discussions the Common Agricultural Policy is
regarded by the Community as non-negotiable. However,
its implications for trade have led to considerable
confrontations with other countries. These have occurred,
for example, in GATT (General Agreement on Tariffs and
Trade), in UNCTAD (United Nations Conference on
Trade and Development), with agricultural suppliers around
the Mediterranean and with countries party to the Lomé
Convention. Other countries desire better access to
Community markets. In general, where such access has
been incompatible with the successful operation of the
Common Agricultural Policy, the Community has not made
worthwhile concessions.

A Formal Analysis

Diagrams 1 and 2 set out the implications for the rest of the world of price decisions made under a policy such as the CAP. The assumptions it makes clearly oversimplify reality but it does draw attention to the effect of Community price policy on international trade. In Diagram 1 the Community supply designated by SH confronts a Community demand designated DH. If trade is allowed to take place supplies can enter the Community at SW, thus the total amount consumed would be equivalent to OQ_2 of which OQ_1 would be produced at home. If a levy is added to the world price bringing the supply price from abroad up to SW + L the quantity consumed within the Community falls to Q_4 while the quantity produced rises to Q_3. The shaded areas marked A and B represent the loss in revenue occurring to overseas suppliers as a result of the operation of the Community's policy.

Diagram 1.

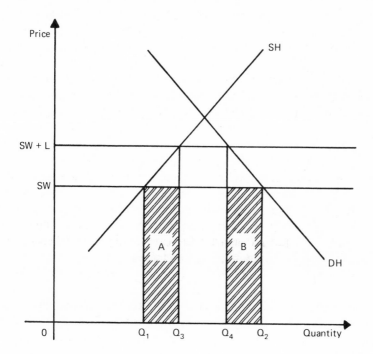

The diagram provides a starting point for discussion of the effects of the CAP. One important element is the position of the world price. Since the Community is the largest trader in agricultural goods in the world, the assumption that the world price is unaffected by its behaviour is unrealistic. As Table 9 shows, the Community's agricultural trade is to an important degree intra-Community. However, it remains a major world trading bloc. Any disturbance in its net imports or exports will have serious consequences for world prices. A second feature of the diagram is the assumption that domestic supply and demand remain in roughly the same position. This is unlikely. Improving agricultural technology implies that at a particular price a higher quantity will be produced. This in effect moves the home supply curve SH farther to the right, diminishing the volume of imports. It is, within a protected situation, somewhat difficult to determine whether the decline in imports owes more to protection or to domestic improvement. Politicians tend to argue the case from either point of view according to the interests of their countries. The demand curve, too, tends to move. In general we can assume that as people become richer, they will tend to buy more food. However, their consumption of food rises rather less rapidly than their consumption of products in general and for some goods, particularly potatoes and bread, consumption may actually fall as people replace 'fillers' with a more varied and attractive diet. For preferred products the full implications of the protective policy adopted by the Community may be offset by a general rise in income levels. Should that improvement in income cease and the rate of increase in domestic production continue, then the pressure on Third country suppliers would undoubtedly grow.

Diagram 2 indicates in schematic form what happens when the degree of protection results in an increase in domestic production in excess of the quantity the Community market is prepared to absorb. In effect from being a net importer of a quantity represented by the difference between OQ_2 and OQ_1 the Community becomes an exporter of a quantity represented by $OQ_4 - OQ_3$. At Community prices the value of these exports is measured by the cross-hatched area. In the world market, however, their value is equivalent only to that part which falls below the line SW. Thus, if the Community is to dispose of its agricultural surpluses in the rest of the world it must pay a subsidy equivalent to the area between SW and SW + L on the volume

Diagram 2.

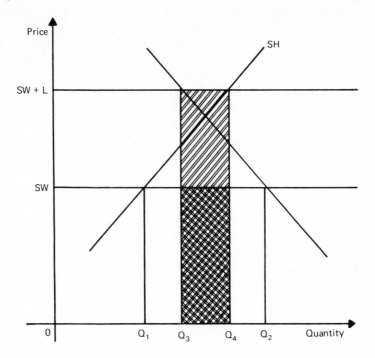

$OQ_4 - OQ_3$. This, from the point of view of the rest of the world, is a clear case of 'dumping'. In reality it is not necessarily more disruptive of world trade than the contraction of Community imports foreseen in Diagram 1: there the displacement of quantities A and B may force traditional suppliers to dispose of exports which formerly went to the EC on the world market at prices well below those prevailing. Selling goods abroad at lower prices than charged internally is regarded as illegitimate within the rules which have governed international trade since the Second World War. However, such rules have not been applied, in the normal way, to agricultural products. Nevertheless, the damage represented by dumping is resented by overseas countries and could form the basis for countervailing action on their part.

The Development of Trade within the Community

Table 10 indicates how the exports and imports of the Community have changed between 1959 and 1976. During this period all trade has increased in value. In general terms agricultural exports from the Community have more or less kept pace with exports of all products. So far as total imports are concerned, these tended to lag behind the growth of exports up to 1974 but have overtaken them in subsequent years. Within this framework agricultural imports have grown less rapidly than imports in general and less rapidly, too, than agricultural exports. The implication is that so far as the rest of the world is concerned the Community has tended to become relatively more self-sufficient in agricultural goods. This feature is confirmed by Table 3 which indicates that between 1959 and 1975 the degree of self-sufficiency in all the major product areas reviewed had actually increased. Taking particular years is never very safe but the evidence reported does correspond to a continuing trend.

It is tempting to conclude that the evidence of greater self-sufficiency proves that the Common Agricultural Policy is working in the way in which the schematic formal analysis in the previous section suggested it must. In fact this is not a valid conclusion. In addition to the Common Agricultural Policy other changes have occurred simultaneously within the Community and independently of the agricultural policy itself. For instance, even if there had been no Common Agricultural Policy, scientific progress in the improvement of plant varieties and animal breeding, better animal feeding and housing, improved fertilisers, and so on, would all have tended to increase domestic supplies. Since, in the period under review, the Community was relatively prosperous compared with the rest of the world, the chances are that these improved methods would have been applied and would have diminished the Community's food requirements from elsewhere. Again, the Community was operating during much of this period in a situation of relatively full employment. It is possible that in the absence of the Common Agricultural Policy the pull of higher incomes in other sectors of the economy might have so much more considerably reduced the size of the agricultural labour force as to result in a diminution in the level of agricultural output. Such hypothetical scenarios cannot be tested objectively by reference to published data. However, they counsel caution in attributing the whole of the effects on international trade in agricultural products, good or bad, to the operation of the Common Agricultural Policy itself.

Arrangements with particular countries

Within the framework of the Rome Treaty members sought to make provision for some less-developed countries which were their dependencies. Articles 131 – 6 allow for trade between these countries and the metropolitan Community. The general principle of this association was that it should 'permit furthering of the interests and prosperity of the inhabitants of the countries concerned in such a manner as to lead them to the economic, social and cultural development they expect' (Article 131). Between 1968 and 1972 eighteen of the African countries which were linked to the Community through these procedures became independent. Their relationships with the Community were renegotiated in terms of new association arrangements embodied in the text of the first Yaoundé Convention which was signed in July 1963 and came into force for a period of five years from 1 June 1964. The terms of association for continuing dependencies were decided by the Council of Ministers. In 1969 a second Yaoundé Convention was negotiated and, at that stage, eighteen members were full associates of the Community.

In 1969 a number of countries in East Africa, Kenya, Uganda and Tanzania negotiated a limited association arrangement with the Community. The advent of the United Kingdom to the Community led to considerable anxieties for those British dependencies which had hitherto enjoyed privileged access to the British market. Since this period coincided with the need to renegotiate existing association agreements, a general discussion took place resulting in the Lomé Convention which applies to former British dependencies in Africa, the Pacific and Caribbean and those African countries which were formally associated with the Community. This Lomé Convention forms the basis of present relationships between many former dependencies and the Community and is itself due for renewal during 1979.

Another route to association with the Community was established through Article 238 of the Treaty of Rome. This permits the Community to conclude an association arrangement with a Third country, with a Union of States or with an international organisation involving reciprocal rights and obligations, joint action and special procedures. Within this framework several Mediterranean countries became associated, at an early date, with the Community, in particular Greece (1961) and Turkey (1963). The arrangements with these countries were undertaken with a view to full membership of the Community some twenty-five years after the agreements were signed. For a further group of countries around the Mediterranean agreements were negotiated in which association was to be achieved at some future date. These included Morocco, Tunisia, Malta and Cyprus. So far as Spain, Israel and Yugoslavia were concerned trade agreements were negotiated but no association had been implied.

The complexity of these arrangements and the tendency of concessions granted to one country to reduce the value of concessions granted elsewhere led to a desire to adopt a global Mediterranean policy. In fact such a policy, although articulated, has never been fully achieved. Since the change of governments in Greece, Spain and Portugal these countries have sought full membership of the Community.

The arrangements under Article 238 gave privileged access to the Community's market for goods produced within associated countries. In return the associate countries were allowed to continue to protect their own industries provided that they did so in a way which did not give inferior terms to goods of Community origin, compared with Third countries. In practice the association arrangements with Greece were frozen for a long period during the regime of the colonels. The end of that period proved to be a prelude for full Greek entry into the Community.

For both the Lomé associates and those closer to Europe the Common Agricultural Policy presents a major hurdle. Access to Europe's food market may be the most important benefit to be gained for countries with large agricultural sectors compared to their industrial activity. However, where imports of food from developing countries conflict with the CAP the Community has not proved very sympathetic. Its most considerable concession was to accept imports of 1·3 m tonnes of sugar, representing the less developed countries' (LDCs') input to the Commonwealth Sugar Agreement under which the United Kingdom regulated its sugar imports before entry. Even here the arrangement has not worked entirely smoothly. For other commodities the arrangements have tended either to limit the quantity which the countries can sell to the EC or to require them to deliver at a price not below that operated within the Community itself. In effect this means they are unable to exploit their comparative advantage in trade with the Community.

So far as the Mediterranean is concerned the admission of Spain, Greece and Portugal will undoubtedly intensify the problems created by the CAP for those countries which remain outside the Community. In so far as the Community makes concessions to these countries it must diminish the economic advantage accruing to its new members. Since the Common Agricultural Policy will itself raise the costs of cereal feeds and meat imported by the countries acceding to the Community, the imbalance within the Common Agricultural Policy may prove increasingly stressful for the Policy itself.

The EEC in GATT

The formal position of the Community has been that the Common Agricultural Policy is an internal matter. However, agricultural issues have figured largely in recent negotiations within GATT. Since the simple dismantling of protection is unacceptable, alternatives have been sought. The most attractive of these has been the negotiation of commodity agreements. In principle the Community appears to favour such arrangements. It has argued that only on the basis of some 'normalised' world market, freed from the excess volatility of traditional world market prices, would it be prepared to adopt a more helpful attitude in terms of access to the Community's market. However, where agreements have been negotiated the Community has not always been an enthusiastic participant. For example, it has not been generous in its approach to the problems of stockpiling cereals nor has it participated in the newly negotiated sugar agreement. Thus the impression remains that the Community operates the Common Agricultural Policy in an autarchic manner and attaches relatively little weight to the implications for Third countries.

5 The Common Agricultural Policy and the Farmer

Many onlookers see agricultural policies as devices to benefit farmers and the agricultural industry. Farmers' unions, co-operatives, the agricultural trade associations and the farming press all seek to persuade governments to improve the lot of agriculture. Political parties and governments have implicitly accepted this role. In policy statements they have endorsed the view that official action can and should help the industry. Farmers and politicians seem agreed on two general objectives, that farmers' incomes should be higher and their prices more stable.

The Rome Treaty adopted a similar but slightly ambiguous stance. It appears to set the improvement of individual earnings as an objective for the CAP. Such a benefit is, however, to be derived from an improvement of the productivity of agriculture (see p. 9 para. 3b of the Treaty). Since the market for farm products is so limited that most are price inelastic, the implication is that earnings will rise only if enough people leave the industry. The undertaking to stabilise markets, Article 39, para. 1c, accords more nearly with the industry's demands but may also be construed as benefiting consumers.

To give effect to these aspects of the Treaty intended to help farmers, the Community has relied on price policy, structural policy and to a small degree aids to producer marketing organisations. Its price policy maintains internal prices at levels judged right for the Community even when these are well above those in world markets. Its structural policy helps to diminish the size of the farm labour force by encouraging early retirement or re-training. Further, through making some capital available to farmers who fulfil certain conditions and submit approved development plans, the Community may help those who remain in farming to earn more. Producer groups can raise farmers' revenues if they are able to regulate the flow of goods to the market; the producer associations formed under the fruit and vegetable market organisation (Regulation 1035/72) and those more recently authorised in Belgium and France under Regulation 1360/78 may be visualised in this light.

To judge whether farmers have benefited from the CAP involves some hypothesis about their situation in its absence. If the alternative were a market open to international competition and consisting of individual, unorganised farmers selling on domestic markets, the CAP can clearly be seen to have helped farmers. In reality such an alternative never existed. In the absence of the Community, member states protected their own farmers to differing degrees and by differing methods. Thus, the relevant question is how do circumstances under the CAP compare with those national governments might have devised? The answers vary from country to country.

The price level initially agreed amongst the Six was a compromise: farmers in West Germany and Italy faced lower prices; in France and the Netherlands prices were raised. Subsequently, the combined effects of a strong currency and oversupplied markets may have tended to keep German prices lower than a German government would have chosen. In France and Britain the opposite seems likely to be true. For the UK, however, a continuation of the system of allowing imports to determine market prices whilst making up farmers' returns by deficiency payment would have been consistent with maintaining farm incomes even at lower market price levels.

Structural policy has remained a partnership between the EC and national governments. Possibly, had the budget costs of surpluses been felt by national exchequers rather than shared among member states, some countries, in particular France, might have done more to assist the out-migration of farmers and farm workers. A report by Professor Vedel[1] in the early 1960s indicated that there was a need to withdraw substantial proportions of both land and manpower from farming in France. Accelerated movement to other sectors might have raised the incomes of those who remained in farming, but the conditions for this involve much more than agriculture and may not be perceptibly affected by the level of expenditure to which either the EC or the member states were prepared to commit themselves. Indeed it is possible that the CAP expenditure in this area has been used principally to reduce national structural expenditure rather than to add to it. On balance it seems improbable that farmers have gained greatly on this score.

Potentially a farmers' marketing organisation is in a strong position. If it can limit supplies to the market the consequence is a sharp rise in revenue. In the UK, producer marketing boards set up under the Agricultural Marketing Acts of 1931 and 1933 enjoyed substantial power. Not all succeeded but those which did, the Milk Marketing Boards, the Hop Marketing Board, the Potato Marketing Board and the Wool Marketing Board, certainly improved the revenues of producers. In other countries other agencies, some with official participation such as l'Office National Interprofessionel des Céréales (ONIC) and others which were farmers' co-operatives enjoying a virtual monopoly, for example, in Holland COVECO (Co-operative Cattle Market and Processing Centre) exercised similar power. It seems unlikely that the CAP's producer associations have an equivalent effect on farmers' incomes. Indeed since the terms of membership are not as specific as co-operatives require, some existing co-operatives have argued that a proliferation of independent groups will dilute producers' market strength. One particular feature of this form of aid to farmers is that it conflicts with the spirit if not the letter of the Treaty in so far as the Treaty embodies rules of competition. Although such rules, under the terms of the agricultural titles, may be set aside where a common market organisation exists, to do so to any very substantial extent is to run the risk of defeating the purpose of creating a Community at all.

Another approach to the effect of the CAP is to examine how farm incomes have evolved over time. To do so involves two sorts of difficulty, conceptual and factual.

The farmer's income is in essence a profit from his business. Calculation of these profits is subject to accountancy conventions. Since these conventions may vary from place to place or from time to time published evidence is not always a very clear guide as to the performance or development of the sector. In arriving at a profit figure allowance has to be made for the opening and closing value of a farmer's assets and the net flow of farm receipts and payments during a particular year. In addition the farmer and his family may themselves consume part of the output of the farm. Since if he did not work on his farm the farmer could hold a paid job, allowance must be made for his manual and managerial input. Very often substantial parts of the farm's capital belong to the farmer himself. Since such capital could, at least in principle, be re-invested in order to earn income in some other enterprise, a failure to charge the farm business for this capital overstates the real profitability of the farm although not the cash at the disposal of the farmer. Thus, what is calculated as a farmer's 'income' depends upon decisions which may be in a degree arbitrary about how labour is charged, about perquisites and about the quantity of capital involved and the rate at which it is to be remunerated.

Since 1958 farmers' incomes in general seem to have risen within the Community. The economic accounts for agriculture[2] 1968 to 1973 show that incomes increased in real terms in every member state. However, since 1973, in France, the Netherlands, Belgium and the UK farmers have, in some years, experienced a decrease in real income. Although it continued to rise in money terms, a more rapid rate of inflation meant that the purchasing power of the farmers' income fell. In addition the commodity boom of the mid-1970s forced up the costs of fuel, fertilisers and feed. Such information cannot confirm or refute the claims of the CAP to have helped farmers. Positive benefits to income can be attributed to the application of improved farming methods which might have occurred in the absence of the CAP. Others may reflect favourable movements in the costs of inputs or the prices of outputs due to economic growth in other sectors, rather than to the activity of the Common Agricultural Policy. Again, the failure of the real incomes of some farmers to rise in the years since 1973 is in part due to unusually severe summer droughts in both 1975 and 1976.

Undoubtedly the most important single reason determining the long-run level of farm incomes is the number of people engaged in the industry. Between 1958 and 1978 the agricultural population of the Six declined from 18 million to 7 million. If this change had not occurred, incomes in 1978 would have been only one-third of their actual level, despite the intervention of the CAP. This complex pattern of income determinants means that simple assertions about the effect of the CAP on farm incomes are unreliable.

One persuasive reason, at least in the eyes of many farmers and agricultural politicians, for improving farmers' incomes is that they compare unfavourably with incomes received by workers in other sectors. However, comparison of farm incomes with industrial incomes is extremely difficult. The income of most workers in the rest of the economy is on a wage or salary basis. Thus, for purposes of comparison, the data available may be measuring rather different concepts in the case of farmers than they are in the case of non-farmers. More complication emerges if an attempt is made to allow for the non-monetary costs and benefits enjoyed by the farming and non-farming communities. For example, the agricultural population may enjoy housing at a lower cost than would be the case in an urban environment, perquisites in terms of food and fuel and the facilities of an open countryside. At the same time rural communities frequently have to pay more for goods delivered from towns, incur higher costs in travelling to towns for business appointments and enjoy a lower standard of social amenities in terms of schools, hospitals, libraries, and so on, than most urban dwellers. Industrial jobs, too, may have perquisites. For example, some employers may provide a company car, others may give access to purchases at reduced prices, others may provide sporting and cultural activities at less than their full cost. Similarly, the industrial worker may incur some costs involved in his work not commonly experienced by farmers. For instance, he may be required to maintain a particular standard of dress, he may need to purchase certain professional books in order to keep abreast of his subject, he may be required to travel considerable distances to his work where a farmer frequently lives 'on the job'. Clearly any very exact balance between farm incomes and the incomes of non-farm workers is impossible. Thus, attempts to measure in money terms the gaps between incomes in various sectors can be misleading.

Table 11, which compares the Gross Value Added per person employed in agriculture, forestry and fishing with GDP per head in the economy as a whole does not measure the gap between farm incomes and others. It does, however, give some insights into the probable development of this relationship. Certainly there is no clear evidence that output per head in agriculture has tended to catch up with GDP per head in those countries where it is lowest – Germany, France and Italy. In the UK and Denmark a relative surge in farm output per head appears to have followed accession but the explanation, apart from price increases, may also lie in the relatively undynamic performance of the industrial sectors of some economies. Thus, although UK farmers seem to have a relatively high output per head compared with the rest of the economy, in absolute money terms their Gross Value Added per head is only

56 per cent of the Belgian, 61 per cent of the Dutch and 114 per cent of the German. Such evidence suggests that income levels may be more strongly related to the success or failure of the economy as a whole than to benefits from agricultural policy.

One source of confusion lies in the distinction between income and wealth. It is frequently said that farmers live poor and die rich. In effect many farmers own assets which, in total value, compare very favourably with the total wealth of most of the non-farming population. However, as long as the farm remains in operation, the land and buildings which it embodies cannot be sold off nor can the machinery used readily be turned into cash. In terms of the UK where land prices tend to be relatively low, a 25-hectares farm would be small. However, at a price of £2,000 per hectare, not uncommon in recent years, the implicit value of the farm land would be £50,000. This is considerably more wealth than most non-agricultural citizens possess but very small compared with that of medium- or large-scale farmers. One curious, somewhat underconsidered aspect of the Common Agricultural Policy is that by raising the price of agricultural products it must tend to accelerate the increase in land prices. As a result farmers who acquire land at current prices may have incomes no higher at CAP prices than would have prevailed at lower price levels but possess a higher-priced capital asset. In the short run the cost of servicing this asset may tend to reduce income. In the longer run, if land prices keep in step with inflation, the farmer's wealth may be greater.

The implication of the CAP for farm incomes has to be viewed in this complex context of changing asset values and varying performances in the rest of the economy. In the long run, the expected level of income represents the supply price of farmers. If it falls too low more farmers will leave the industry, voluntarily or involuntarily, and fewer new recruits will enter. If prospects seem favourable more will join or remain. The outcome of these adjustments is that incomes tend to move in the direction of an equilibrium 'price for farmers'. If the CAP, through raising producer prices, enhances farmers' expectations, the chances are not that income will rise but that there will be a larger number of farmers.

It seems that the prices maintained through the CAP must have raised farmers' expectations of income compared with what they would have been had the EC market been accessible to imports from the world. It may not have seemed to offer prospects as good as those which would have existed in some member countries, notably Germany, in the absence of the Community, on the assumption that national governments would have had their own policies. For the agricultural exporting members, however, the CAP has probably sustained higher prices and allowed more farmers to continue in business than national governments would have been willing to finance.

The Community has taken considerable pride in the stabilising effect of its policies on agricultural and food prices. At first sight this must seem beneficial to farmers. However, it is worth noticing that there is an inconsistency between the general argument and the case of any particular farmer. The Agricultural Policy as it stands raises the revenues of farmers by increasing the prices of the goods they sell. However, an important part of the goods sold by farmers are in fact inputs to other agricultural enterprises. Thus, if cereals are particularly abundant stabilisation prevents prices falling; this means that livestock farmers do not have access to cheap feed. This lack of homogeneity of interest is a matter of considerable importance in assessing whether the degree of stability and the level of stable prices chosen for the Common Agricultural Policy have proved helpful. The level of price fixed has resulted in the accumulation of stocks of some products which cannot be sold at EC prices within the Community and which ultimately may have to be exported with the aid of substantial subsidies. It is not clear, since the proportion of livestock output compared with arable production is tending to grow, that this policy which keeps feed costs high has necessarily provided the best form of stabilisation for farmers as a whole.

The claims of the policy to have improved farmers' incomes and stabilised their markets have at least some evidence to support them. There is, unfortunately, also considerable evidence that in pursuit of these goals some very serious problems remain unresolved. Four of these require special mention.

First, the geographic distribution of benefit in terms of farm incomes is not satisfactory. The Common Agricultural Policy is principally a policy of high prices. This means that those who produce most derive greatest benefit from the policy. However, in less favoured regions, where incomes tend to be lowest, each farmer produces relatively little. Thus the CAP tends to exacerbate rather than relieve regional income disparities. It is

thus understandable and welcome that towards the end of the 1970s the Community's emphasis in agricultural policy has swung to a greater degree of regional discrimination. Even this may not represent a very sensible or satisfactory approach. If regional discrimination in the form of even higher regional product prices encourages a relative growth of agricultural activity in areas which are basically uncompetitive, continuing aid will be needed on a long-term basis if production in such vulnerable regions is to remain profitable. Such arrangements may contain the problem in the short run; they do not represent a solution to it.

The price orientation of the policy which leads to its geographic inequity also leads to inequity between small and large farmers. In general small farmers, who have less to sell, must be expected to have poorer incomes than larger farmers. Because they may only be able to expand by acquiring more agricultural inputs, both stock and feed, the effects of a policy which raises the costs of inputs is to reduce the size to which they can increase their businesses given their limited access to capital. Thus the CAP may be more damaging than appears at first sight. It is important to note, in order to gain an appropriate perspective, that not all small farms are the sole source of income for the farmer and his household. Increasingly in Europe smaller farms have become either part-time or spare-time ('hobby') farms. The precise degree of dependence upon the farm varies infinitely from family to family. However, it is clear that at least as a transitional stage part-time farming may achieve a redistribution of labour in a more effective and painless way than increases in prices paid for agricultural products. The Common Agricultural Policy has done nothing to encourage such part-time farming.

A third feature of the policy which is unsatisfactory as far as some of the Community's farmers are concerned is that it is, on the whole, more favourable towards products produced in the north than in the south. This reflects the different product mix, principally livestock and cereals in the north as against fruit, vegetables and wine in the south. In terms of equity this is unfortunate. Many of the farmers in the south have very low incomes and depend heavily upon the products which claim least support.

Although the incomes of farmers in the EC have improved since 1958, it can be argued that this is not simply the result of the

Common Agricultural Policy but rather because of the sustained outward migration of workers from the industry. Those who misread the undertakings of the Rome Treaty to farmers as being to raise the incomes of all those farming in 1958 may feel that the policy has failed. However, the CAP has not prevented changes which have collectively raised the incomes of those who remain in farming.

This chapter started by noting that many regarded agricultural policy as being devised for farmers. Such a concept is misleading and may at times be dangerous. Agricultural policy exists for the Community as a whole. In doing so it must pay attention not only to the concerns of farmers but also to the problems of member states, the problems of consumers and the interaction of agriculture within the economy of the Community as a whole. The most serious criticism which might well be levelled at the Common Agricultural Policy is that it has consciously directed expenditure towards a short-run 'holding' activity rather than attempting to promote a more dynamic and positive approach to long-run adjustment problems. Its contribution in terms of encouraging retirement and re-training represents a recognition of the need for such a long-run approach. However, expenditure under these headings has been trivial compared with the cost of price supports. Thus, it is difficult to conclude that a policy which necessarily must be constrained by interests other than those of farmers has, despite considerable expenditure, done as much to raise the real incomes of those who work in agriculture as it might have done had it sought more bravely to be an agent of adjustment in a changing and developing European economy.

Notes: Chapter 5

1   *Rapport général de la Commission sur l'avenir à long terme de l'agriculture française presidée par M. le doyen Georges Vedel* (1979). For a detailed outline of the report, see Professor A. Bienayme, 'The Vedel Report proposals for reform of French agriculture', in Federal Trust, *Current Agricultural Proposals for Europe* (Conference Report, 1970).

2   *Source*:   EC Commission, *The Agricultural Situation in the Community 1978 Report*. 'Income is measured by the net value added per person employed in agriculture. It represents the return of inputs to agriculture.'

6 The Common Agricultural Policy and the Consumer

The Rome Treaty promises four things to consumers. First, that agriculture should be efficient. Secondly, that it shall ensure a secure supply of foodstuffs. Thirdly, that prices will be stable. Fourthly, that they will be reasonable. In practice agricultural policy in the context of the Community has often seemed to give much more weight to the needs and the problems of farmers than to the implications of the policy for consumers. It seems to have been accepted that a prior goal of agricultural policy was to aid farmers. In the United Kingdom this doctrine has been strongly challenged by consumers. There the grounds of concern are related to a desire both to keep food prices down and to minimise the cost to the UK economy of the CAP. It is still accepted that the policy should help farmers. Consumers can therefore expect little when they examine the performance of the policy in the context of the four promises made by the Treaty.

At first sight, neither price policy, which maintains a level of prices consistently higher than that needed to ensure an adequate flow of foodstuffs, nor structural policy, which artificially reduces the price at which certain inputs may be incorporated within the agricultural sector, is conducive to economic efficiency measured in the classical sense. Before such a judgement is fully accepted, however, it is prudent to examine the possibility that market prices imperfectly reflect the true value of resources devoted to agricultural production. This certainly would be the view of those who defend farm support on the grounds that agricultural activity of a traditional character maintains a desirable countryside and rural way of life. Essentially their argument amounts to an assertion that the output of the farming industry is not simply food and industrial raw materials but also a service from which many consumers benefit but for which very few pay directly. Similar arguments may be used to justify structural schemes which support particular techniques, new or old, in order to preserve a pattern of rural activity on a smaller scale than might emerge from the unmitigated operation of market forces.

Studies of such costs and benefits have demonstrated the difficulty of quantifying non-market aspects of the value of the resources used or the outputs produced by an industry. In fact, little attempt has been made to justify particular expenditures under the CAP on the grounds that it would add to overall efficiency in this way,[1] although the more regional aspects of structural policy, discussed in Chapter 3, fall into this category.

If the narrower definition of agricultural output, sales of food and raw materials, is employed it is clear that in so far as the CAP retains a higher volume of resources within farming than would otherwise be the case consumers must suffer. Losses occur in two ways: first, because the prices they pay, under the arrangements of the CAP, are higher than those needed to ensure satisfactory supplies; secondly, because some of the resources retained within the industry could produce goods of greater value to consumers should they be redeployed elsewhere. The mechanisms of the CAP which maintain higher levels of consumer prices as a means of supporting producers are particularly damaging. In contrast a deficiency payment scheme, such as that which operated within the United Kingdom in advance of membership of the Community and which the EC operates for oilseeds, has much to commend it because the market is allowed to clear. Prices fall in those situations in which aggregate supply is rising relative to aggregate demand whilst producers are protected through supplementary payments. Thus the visible embarrassment of accumulated surpluses is avoided and consumers do not have to pay higher prices than are neeeded to ensure adequate supplies. Indeed, since the overall effect on production may be to raise the level of output consumers may enjoy food at lower prices than would otherwise have been possible. The inapplicability of deficiency payments on the scale of the Community, in so far as its principal agricultural products are concerned, is one reason for continuing consumer dissatisfaction. Inapplicability occurs, first, because of the undoubted administrative complexity of a deficiency payment scheme within the context of a Community, and secondly, because of its probable financial consequences in terms of calls on the Community's budget.

Deficiency payments transfer the cost of agricultural support from consumers to taxpayers. The size of the budgetary commitment depends upon the gap between the price guaranteed to producers and that which applies in the market and upon the volume of output produced domestically. A deficiency payment system would allow market prices to fall when supplies on world markets were more plentiful. If the EC, in such circumstances, had retained prices to producers close to their existing levels the cost per unit of farm output would have been very high. For example, EC butter prices have at times been five times the prevailing world price. Thus, for every pound spent by consumers on butter, the Community taxpayers would have had to spend four pounds. Again, the Community is almost self-sufficient in most temperate products and more than self-sufficient in some. Compared with the UK where only 60 per cent of indigenous-type goods were produced at home, the quantity of farm output to be supported would have been very much higher. For a Community with limited budgetary resources such a policy could not be applied to the bulk of farm production.

The regressive effect of the CAP support system which makes food prices higher has attracted considerable attention within the United Kingdom. People who are poor spend a higher proportion of their income on foodstuffs than wealthier citizens. A policy which increases the price of food represents a higher proportional tax on the expenditure of the poor, even though the total amount of tax collected per head will be higher from rich families than from poorer families. Thus, the CAP has an adverse effect on the distribution of real incomes.[2]

The arguments considered so far are concerned with the static effects of the CAP: the difference it creates in a given situation. One possible line of defence might be that its dynamic effects offset these disadvantages. It might be argued that by accepting this level and form of protection the Community is able to ensure a higher rate of growth in real income for all citizens. As a result, although the poor may spend a rather larger share of their income on food than would be the case should a more liberal policy be pursued, nevertheless the level of their income is so much higher that their total ability to buy food is enhanced. To validate this argument requires some analysis of the role of agriculture within the economy of the various member countries of the Community. For example, if it could be demonstrated that some particular country was unable to pursue a rapid rate of growth because it ran into severe balance-of-payments problems and that the most effective way of reducing those balance-of-payments problems was to increase the domestic production of food, then a policy like the CAP which increased output and diminished imports might contribute to economic growth. It is worth noting such dynamic arguments need not always prove favourable to the CAP. For instance, if the view is adopted that the primary constraint on growth is a tendency for inflation to accelerate to a level at which public action to control the price level frustrates the growth of the economy, then the CAP may be held to have a negative effect. This stems from a belief that higher food prices will be reflected in wage demands, and that wage demands will be conceded and passed on to consumers in the form of higher prices, in turn leading to further claims for yet higher wages and for pressures to re-assert existing income differentials. The chain of argument is long and contentious but if it is accepted then by increasing food costs the CAP may be held to restrain the rate of growth rather than to accelerate it.

The apologists for the CAP frequently point to its success in maintaining secure supplies of food. They indicate, not without some justification, that whilst consumers are ready to complain vigorously about the anomalies of large surpluses of butter, wine or cereals, they complain even more strongly if supplies fall short of accustomed levels. Thus, the argument runs, the policy, by stimulating a higher level of domestic output, enhances the food security of the Community. Not all who are concerned with consumer interests share this view. For example the UK Consumer Association has in its publications sought to distinguish between those products for which regularity in supply is very important and those for which it matters relatively little. Clearly, where ready substitutes exist for a particular product, regularity in supply may be of less significance than for products such as potatoes or cereals where no near alternative, in the eyes of consumers, is to hand. For the first group security of supply is clearly less significant than for the second.

Even for products for which security is important the CAP may not be the best way to ensure it. A different way to ensure supplies is to accumulate reserve stocks. Another approach is to diversify the sources of supply by relying on trade and entering into agreements with overseas suppliers. In general it is arguable that the likelihood of a failure of production in all the various major areas of the world is less than failure of production in one particular zone, even if that is, as is the case in Europe, one of the more favourable climates of the world. The approach of the CAP is to stimulate a higher volume of domestic output. This certainly ensures, when it succeeds, adequacy if not superabundance in food supplies. However, marginal increases in domestic output frequently rely upon imports of various types of input. Two deserve special mention. Livestock production can most readily be expanded by importing more livestock feed, in particular, soya. Expanded crop production relies upon the use of considerable quantities of fertilisers, of sprays and of machinery. Each of these is dependent upon an adequate supply of fuel, principally oil. Thus, the security problem is transferred from one of food to one of the means of producing food. It is by no means clear that the Community can be more assured of its supply of soya or oil than it could be of a supply of food ready for consumption. Analysis of such questions depends very much upon political judgement. For instance, those who advocate international co-operation as a solution may be unduly optimistic about the chances of the varying nations of the world agreeing to regulations which at any one time appear disadvantageous to some of the countries concerned. Those who advocate greater levels of domestic output are indicating confidence about the future availability of imported supplies of raw materials. Such judgements are implicit in the attitudes taken to food security by the CAP but seem to be relatively little discussed.

As with security so with stability of price the CAP appears to have done well. By comparison with both world prices and the internal prices in non-Community countries it seems to have succeeded in restricting price movements. However, even here its achievements have not been accepted without some criticism. For instance, stability has, as was pointed out in Chapter 4, been purchased to some extent at least at the expense of the rest of the world. Abundance within the Community depresses world prices but not the EC price. Shortages in the EC are offset by imports which may tend to exacerbate world scarcities. From a purely consumer point of view stability may be purchased at too high a price. The maintenance of lower prices in the years when crops do badly has to be set against the higher prices operating at other times. If high-priced years are infrequent the cumulative cost may offset the occasional benefit. In fact it might be less costly for consumers or governments to hold a suitable stock of money in order to cope with the odd year of high prices than to accept higher internal prices and the consequent resource misallocation.

Probably the most elusive of the concepts embodied in the Treaty of Rome, so far as the CAP is concerned, is that of 'reasonable prices'. Almost any price level can be made to seem reasonable to someone. Thus, farmers frequently feel that prices ought to reward 'efficient' producers, efficiency in such a context usually being conceived in technical terms such as yield per acre. Equally, consumers may feel it is unreasonable to pay any more for a product than is necessary to ensure that they can buy the product when they want it.

In practice reasonableness very often seems to relate to relativities over time and between income groups. For example, a price level which is in terms of world prices high may seem more reasonable to German consumers whose food costs may, in real terms, be falling than it does to British consumers to whom it represents a substantial increase in living costs compared with past levels of price. Again, farmers may feel that it is unreasonable that wages should go up by certain percentages each year whilst agricultural prices are, in their own words, 'held down'.

This imprecision in the concept of reasonableness makes it extremely hard to appraise the performance of the policy. Even amongst consumers very different attitudes may coexist. One important feature, however, must be noted. The food price paid by consumers embodies services of distribution and processing well beyond the farm gate. Indeed, in many countries more is now spent on the activities which occur after the farm than is spent directly on the products of agriculture itself. Thus, movements in CAP prices which generally refer to wholesale markets for agricultural products may be attenuated before their affect on consumer prices is noticed. Again since distributive and processing activities in the food industries tend to be relatively labour-intensive and the rate of wages amongst the member countries of the Community varies considerably, the implications of a particular change in farmgate prices for food consumers may be rather different.

Finally, it is worth stressing that the basket of goods purchased by consumers within the Community countries varies. Thus, a price shift which may seem of relatively little significance to consumers who buy small quantities of the product may loom much larger in the minds of those for whom it is an important element in diet. The British obsession with butter may be psychological as much as economic. However, their concern with livestock products relative to fruit and vegetables reflects a different pattern of consumption to, for example, that of Italy. Overall it seems fair to say that the consumers are reluctant accepters of the CAP as it stands and believe in the words of their representative organisation, BEUC, that the Community's policy should shift towards being a 'food policy' rather than simply an agricultural policy.

Notes: Chapter 6

1   For a fuller discussion of some of the problems involved in such types of appraisal the reader is referred to M. C. Whitby and K. G. Wells, *Rural Resource Development: An Economic Approach* (1974).

2   This issue is well discussed in T. Josling and D. Hamway, *Burdens and Benefits of Farm-Support Policies*, Trade Policy Research Centre No. 1 (1972).

7 The Common Agricultural Policy and the Community

Earlier chapters have discussed the reasons for the existence of the Common Agricultural Policy, its principal features and its implications for other countries, farmers and consumers. In this final, brief chapter the role of the policy in the Community is discussed.

Probably the most important achievement of the CAP is that without it there would have been no Community. Discussions in the 1950s made it clear that some potential members, notably France, saw a secure market for their agricultural exports as an essential feature of any European arrangements for economic unity. The failure of the contemporary British Free Trade Area proposals was in part due to a conscious British decision to exclude agriculture. Within the Six, the agricultural policy was seen both as a *quid pro quo* for agricultural exporters for the gradual dismantling of industrial protection and as a means of tackling together fundamental problems of economic adjustment and trade which faced all agricultural policy-makers.

But if the existence of a common policy for agriculture was essential, then its character was no less essential. A policy of high prices and rigorous protection from Third country competition enabled the Community at once to ease the burden of high-cost producers in Germany and to ensure substantial balance-of-payments benefits to France. Common financial responsibility meant that even if at these prices no markets existed in the Community or abroad, all member countries would have to bear a substantial share of the cost of exporting to world markets. Thus the scene was set for agricultural expansion whilst those who feared competition from lower-cost agriculture were safeguarded.

In operation the CAP has assumed both a symbolic and a financial importance far larger than the role of farming in the economy of the Six. It absorbed most of the Community budget. It demanded the energy and skill of many 'Eurocrats'. It became a focus of debate and evidence of the Community in action. At the same time various interests, both national and sectoral, became entrenched in its operation. Thus the agricultural exporters sought to safeguard the benefits they had secured, whilst milk, cereal and sugar producers sought to retain their right to expand production without penalty in terms of lower prices. So convinced became many members of the Community of the vital importance of the CAP that to criticise it, or even to propose its modification, was sometimes regarded as disloyalty to the idea of the Community itself.

Such an attitude is not tenable today. Five factors have combined to challenge the established character of the CAP: nationalism, technical progress, changed social priorities, enlargement and economic recession. Each of these needs further explanation.

The Community was founded on the principle of a pooling of sovereignty in economic affairs. At the outset each member retained a virtual right of veto but ultimately this was intended to be replaced by a complicated system of decisions by qualified majority. In practice the member governments have withheld their acceptance of qualified majority decisions on matters of vital national interest. Agricultural price-fixing and policies are regarded as of such vital interest that unanimity has continued to be the rule. This means that major decisions must be a compromise of national interests rather than an expression of a common interest. The initiative, although formally remaining with the Commission, has in fact moved to the Council. In agriculture this has led to stalemate. Prices cannot be cut, so there is a ratchet process of price increases. Sometimes these fall below the rate of inflation, allowing a real price reduction to occur, but the Community has lost the ability to move prices decisively in response to long-run market developments. Similarly, Community initiatives in regional and structural policy are discussed at length but inadequately financed and even then on a basis which means that the net transfer across Community frontiers tends to be small.

Technical progress has condemned a Community saddled with so inflexible a price policy to growing problems of surplus. By improving yields and reducing costs farmers offer increasing output to a static or slowly growing market. The CAP and national governments have accentuated the problem by providing aids designed to accelerate the increase in farm productivity. Ultimately, if no change is made in the policy and no addition accrues to the 'own resources' of the Community, the cost of financing surpluses will exceed the funds available to the budget.

Social priorities inevitably and rightly complicate the operation of agricultural policy. However, price policy is an inefficient way of raising the incomes of poor farmers or helping backward regions. Increasingly, demands for more positive structural intervention, more aid for less favoured areas and more regionally directed policies demonstrate an awareness that many of the most acute areas of need, personal and regional, have not been greatly helped by the CAP. More recently there are pressures, at least in some parts of the Community, to give greater consideration to other social priorities. Poor consumers also deserve attention. Consumer organisations have argued that the CAP should become a 'food' policy. Rural life involves much more than farming. Environmental and ecological aspects of agricultural policy have attracted growing attention. The countryside is a place of recreation as well as an agricultural workshop; hence some wish to see the CAP develop into a comprehensive rural policy.

In 1973 enlargement of the EC was approached as a gradual adaptation by the new members, Denmark, Ireland and the UK to policies, including the CAP, already operating among the Six. In practice, and in part due to an unprecedented rate of inflation, this transition proved far from smooth. Fundamentally there was a failure to realise that an enlarged Community was a new Community in which compromises acceptable to the Six would prove unworkable among the Nine. In particular the economic situation and traditions of the UK made the application of the CAP more onerous than for any other member.

By exploiting MCAs the approximation of UK prices to the common level was delayed. However, at the same time the determination of the value of the unit of account in terms of strong Community currencies meant that the real level of common prices was continually being raised. This was inappropriate for the Community, especially for the UK, confronting inflation and farm surplus. The failure to adjust to the circumstances of enlargement meant that, as the transitional period ended, the UK, one of the poorest EC countries, found itself a net contributor on a scale in excess of £1,000m. to the Community's operations. Inevitably this created disillusionment in Britain and resentful unwillingness to accept any extension of the Community's activities.

Perhaps the most serious challenge to the CAP arises from the changed European economic environment since the oil crisis and the commodity boom of the mid-1970s. Several elements are significant. First, there are no longer good prospects in other industries for people displaced from farming. Secondly, the markets for farm output now grow more slowly in response to the slower rate of income growth, accentuating the embarrassment of rapid technical advance. Thirdly, whilst real incomes were rising, higher food costs could be absorbed without undue damage to consumers. In a situation of static or falling real incomes this is not true.

The economies of the Nine have coped with very varying degrees of success with the need for economic adjustment. Some have managed to maintain relatively low rates of inflation and unemployment. Others, the less successful, have experienced inflation at times approaching, if not exceeding, 20 per cent per annum and considerable unemployment. Agricultural policy touches upon these matters in terms of both prices and the maintenance of jobs. To link together the farm sectors of national economies without creating greater harmony in the rest of their economies places strains upon national governments and the common policy. The price which seems high to British workers may seem intolerably low to German farmers. The expansion of butter production may seem a means of exploiting the natural advantages of Ireland and the West of the UK at common prices; for the Community it spells growing financial burdens in relation to dairy policy. In the end the members must either achieve a broader degree of economic policy determination at a Community level or reconcile themselves to a less common agricultural policy.

This chapter started by asserting that the CAP was vital to the creation of the Community. It must end by arguing that a common policy is still essential but that the condition of its survival must be a readiness to respond to changing economic, technical and political circumstances. Many useful ideas for reform have been canvassed over recent years. It is beyond the scope of this book to analyse or compare such proposals. It is not, however, beyond its remit to note that if change does not occur, not only the CAP but the Community itself may be in peril.

Table I(a) Numbers of persons engaged in agriculture (thousands)

Year	W. Germany	France	Italy	Neth.	Belgium	Lux.	UK	Ireland	Denmark	EEC 6	EEC 9
1955	4,285	5,041	7,740	533	310	26	1,171	442	505	17,935	
1958	3,978	4,453	6,974	495	276	24	1,061	407	475	16,200	
1965	2,966	3,538	4,956	388	230	19	850	340	385	12,097	
1968	2,523	3,114	4,418	352	201	13	853	310	291		12,075
1977	1,656	2,022	3,149	289	123	9	661	236	218		8,363

Sources: 1955–65 figures: OECD, *Agricultural Statistics, 1955–1968*.
1968, 1977 figures: EC Commission, *The Agricultural Situation in the Community 1978 Report*.

Table I(b) Proportion of population engaged in agriculture (as % of total civilian employment)

Year	W. Germany	France	Italy	Neth.	Belgium	Lux.	UK	Ireland	Denmark	EEC 6	EEC 9
1955	18·5	26·9	40·0	13·2	9·3	19·4	4·6	38·9	24·9	26·1	
1960	14·0	22·4	32·8	11·5	7·6	16·4	4·3	37·3	21·2	20·9	
1965	11·1	18·2	26·1	8·8	6·4	13·7	3·4	32·0	17·0	16·5	
1968	10·6	16·6	24·1	8·3	5·8	13·0	3·1	30·5	16·6	15·4	
1976	7·0	10·9	15·5	na*	3·4	6·0	2·7	23·8	9·3		8·4

*na: figure not available.
Sources: 1955, 1960, 1965 figures: OECD, *Agricultural Statistics 1955–1968*.
1968, 1976 figures: EUROSTAT, *Yearbook of Agricultural Statistics, 1978*.

Table 2   Gross National Product derived from agriculture (%)*

Year	W. Germany	France	Italy	Neth.	Belgium	Lux.	UK	Ireland	Denmark	EEC 6	EEC 9
1956	7·5	10·2	19·7	10·7	7·3	9·0	4·5	27·2	18·5	10·7	
1957	7·2	10·1	18·4	11·0	7·7	8·4	4·5	29·1	17·0	10·3	
1958	7·1	10·6	19·0	11·2	7·4	8·8	4·3	26·0	16·0	10·6	
1965	4·4	7·7	13·3	8·3	6·3	6·3	3·4	21·1	11·0	7·4	
1966	4·2	7·4	12·5	7·4	5·7	na	3·3	19·6	10·2	7·0	
1967	4·1	na	12·4	7·2	5·6	na	3·3	19·5	9·4	6·8	
1975	2·5	4·9	8·6	4·4	2·8	3·3	2·5	15·4	5·5		4·2
1976	2·8	5·2	8·6	4·9	3·2	3·3	2·6	16·5	5·3		4·4
1977	2·6	5·0	8·4	na	2·6	na	na	18·2	5·7		na

* 1955–67 data refer to Gross Domestic Product at factor cost in national currencies and at current prices (include forestry, hunting, and fishing); 1975–77 data refer to Gross National Product at factor cost and at current prices (exclude forestry and fisheries).

Sources: 1955–67 figures: OECD, Agricultural Statistics 1955–1968.
1975–77 figures: EUROSTAT, National Accounts, 1978.

Table 3 Degree of self-sufficiency (%)

	W. Germany	France	Italy	Neth.	Belg./Lux.	UK	Ireland	Denmark	EEC 6	EEC 9
Wheat										
1959/60	74	113	87	35	69	43	na	88	90	
1966/67	84	134	97	63	67	47	na	91	104	
1976/77	95	188	92	56	64	60	50	142		106
All cereals										
1959/60	74	101	98	36	52	na	na	na	84	
1966/67	72	120	68	33	38	na	na	na	81	
1976/77	80	153	71	25	39	65	68	104		88
Sugar										
1959/60	89	122	100	113	113	30	na	141	104	
1966/67	85	89	104	103	117	36	na	123	94	
1976/77	108	158	82	136	173	26	113	156		98

Milk and dairy products (fat basis)

1960/61	92	106	96	138	97	55	na	239	103
1966/67	99	110	91	145	104	53	na	197	106
1976/77	102	101	94	100	100	100	100	101	100
Butter									
1960/61	93	114	74	157	109	10	na	326	103
1966/67	101	114	77	210	108	6	na	330	109
1976/77	132	111	58	390	101	20	225	322	104
Beef and Veal									
1959/60	86	109	76	107	95	70	na	310	95
1966/67	88	109	67	107	87	73	na	269	88
1976/77	95	113	60	130	93	77	575	283	99
Pigmeat									
1959/60	94	100	95	159	107	60	na	332	100
1966/67	94	90	83	187	125	53	na	478	99
1976/77	87	85	74	212	176	64	137	364	99

*Milk and dairy products 1960/61 and 1966/67 figures not strictly comparable to 1976/77 figures.

Sources: 1960/61 and 1966/67 figures: Ministry of Agriculture and Fisheries of the Netherlands, *Selected Agri-Figures of the EEC*, 1970.
1976/77 figures: EUROSTAT, *Yearbook of Agricultural Statistics*, 1978.

Table 4(a) Number of agricultural holdings of 1 hectare and over by size group (thousands)

Year	Hectares	W Germany	France	Italy	Neth.	Belg.	Lux.	UK	Ireland	Denmark	EEC (6)
1960	1 under 5	617·4	464·3	1,880·0	87·7	96·3	3·3	130·5	56·9	34·8	3,372·0
	5 under 10	343·0	375·8	525·0	62·2	52·7	1·9	58·4	65·8	54·3	1,539·0
	10 under 20	286·5	472·7	230·0	53·9	35·2	2·7	72·1	83·5	54·5	1,290·0
	20 under 50	122·0	362·9	88·2	24·5	12·3	2·3	99·4	57·6	43·7	813·0
	50 and over	16·3	97·8	33·1	2·0	2·2	0·2	82·7	14·7	6·4	255·7
	Total	1,385·3	1,773·5	2,756·3	230·3	198·7	10·4	443·1	278·5	193·7	7,269·7
1967	1 under 5	487·5	374·9	1,700·7	70·5	54·7	1·9	118·0	55·7	21·8	2,886·0
	5 under 10	271·8	306·9	450·2	49·2	39·4	1·2	50·6	59·5	37·5	1,266·0
	10 under 20	288·6	413·2	207·4	55·4	35·2	1·9	59·7	82·5	51·0	1,195·0
	20 under 50	141·0	371·8	84·4	25·9	15·5	2·5	86·1	58·5	44·4	830·0
	50 and over	17·4	109·1	34·6	2·2	2·4	0·3	78·7	14·8	7·8	267·4
	Total	1,206·3	1,575·9	2,477·3	203·2	147·2	7·9	393·0	271·0	162·6	6,444·0
1973	1 under 5	344·5	280·0	na	37·5	35·3	1·3	46·6	na	16·7	na
	5 under 10	195·0	208·0	na	33·0	26·5	0·8	36·2	na	26·9	na
	10 under 20	231·0	310·0	na	46·5	30·6	1·2	45·5	na	39·5	na
	20 under 50	173·5	365·0	na	29·6	18·5	2·5	75·5	na	43·1	na
	50 and over	23·8	137·0	na	3·0	3·1	0·4	82·8	na	9·7	na
	Total	967·8	1,300·0	na	149·6	113·9	6·1	286·6	na	135·9	na
1977	1 under 5	287·5	225·0	na	33·0	29·0	1·0	35·9	na	14·4	na
	5 under 10	165·7	175·0	na	28·9	20·9	0·6	32·9	na	23·2	na
	10 under 20	199·4	250·0	na	41·2	26·7	0·9	41·4	na	34·3	na
	20 under 50	178·9	355·0	na	30·4	19·1	2·1	70·1	na	41·8	na
	50 and over	28·3	143·0	na	3·5	3·6	0·6	81·4	na	10·7	na
	Total	858·7	1,148·0	na	137·0	99·3	5·2	261·8	na	124·4	na

Source: EUROSTAT, Yearbook of Agricultural Statistics, 1978.

Table 4(b)   Proportion of holdings in each size group (%)

Year	Hectares	W. Germany	France	Italy	Neth.	Belg.	Lux.	UK	Ireland	Denmark	EEC (6)
1960	1 under 5	45	26	68	38	48·5	32	29·5	20	18	46
	5 under 10	25	21	19	27	26·5	18	13	24	28	21
	10 under 20	21	27	8·5	23	18	26	16	30	28	18
	20 under 50	8	21	3	11	6	22	22·5	21	23	11
	50 and over	1	5	1·5	1	1	2	19	5	3	4
1967	1 under 5	40	24	69	35	37	24	30	21	13	48
	5 under 10	23	20	18	24	27	16	13	22	23	20
	10 under 20	24	26	8	27	24	25	15	30	31·5	18
	20 under 50	12	24	3	13	10	32	22	22	27·5	11
	50 and over	1	6	2	1	2	3	20	5	5	3
1973	1 under 5	36	22	na	25	31	21	16	na	12	
	5 under 10	20	16	na	22	23	13	13	na	20	
	10 under 20	24	24	na	31	27	20	16	na	29	
	20 under 50	18	28	na	20	16	41	26	na	32	
	50 and over	2	10	na	2	3	7	29	na	7	
1977	1 under 5	34	20	na	24	29	19	13	na	11	
	5 under 10	19	15	na	21	21	12	13	na	19	
	10 under 20	23	22	na	30	27	17	16	na	27	
	20 under 50	21	31	na	22	19	40	27	na	34	
	50 and over	3	12	na	3	4	12	31	na	9	

Source: Figures derived from EUROSTAT, *Yearbook of Agricultural Statistics*, 1978.

Table 5 *Capital committed to agriculture tractors (h.p. per 100 ha of agricultural area in use)*

Year	W. Germany	France	Italy	Neth.	Belgium	Lux.	UK*	Ireland*	Denmark*	EEC (9)
1958	92	44	32	61	54	115	na	na	na	na
1965	188	88	76	144	121	157	76	38	193	101
1973	329	160	178	276	248	254	84	78	254	175
1977	399	204	237	349	300	286	122	104	300	221

*UK, Ireland, Denmark figures: estimates by EUROSTAT.
Source: EUROSTAT, *Yearbook of Agricultural Statistics, 1978.*

Table 6 *Products as proportion of value of final agricultural production in each member state and in the community as a whole, 1977 (%)*

	W. Germany	France	Italy	Neth.	Belg.	Lux.	UK	Ireland	Denmark	EEC (9)
Wheat	3·7	8·5	6·0	1·3	3·0	1·8	5·8	1·6	2·2	5·3
Sugarbeet	3·1	2·8	2·0	2·5	3·9	na	2·0	2·1	2·7	2·6
Milk	22·5	16·5	12·8	26·6	15·7	37·2	22·1	30·4	25·6	19·5
Beef and veal	16·9	16·6	10·8	13·1	17·0	28·4	15·1	37·5	13·8	15·5
Fruit and veg.	2·4	4·9	11·0	3·4	4·0	1·9	2·2	1·0	1·0	na
Table wine	0·1	3·5	4·8	0·0	0·0	0·4	0·0	0·0	0·0	1·9

Source: EUROSTAT, *Agricultural Accounts, 1978*

Table 7   Prices of some major agricultural products (Deutschmarks/100 kg)

	W. Germany	France	Italy	Neth.	Belgium	UK	Ireland	Denmark
Wheat	Weighted average price received by farmers for all types and sales							
1957/58	42	29·4	48·6	30·6	39·2	32·5	33·5	27·4
1967/68	38·6	37·3	44·9	39·6	38·2	24·2	30·4	27·0
1975/76	48·3	38·3	49·2	41·2	40·8	34·5	36·0	na
Cattle	Weighted average price received by farmers liveweight for all types and grades							
1957/58	172·1	na	226·2	186·8	173·6	158·1	142·8	142·9
1967/68	232·5	247·0	276·7	na	249·1	166·2	135·8	140·8
1975/76	344·7	313·2	381·7	322·9	319·8	227·2	213·6	284·4
Pigs	Weighted average price received by farmers liveweight for all grades							
1957/58	219·7	na	227·6	181·3	190·8	217·7	194·4	231·7
1967/68	226·1	233·6	279·7	227·6	224·7	180·3	188·2	246·9
1975/76	343·0	397·0	304·6	313·1	293·6	250·8	254·6	337·0
Sugarbeet	Average price realised over whole harvest							
1957/58	7·1	5·3	5·9	5·5	6·1	7·0	7·9	5·9
1967/68	7·6	5·2	6·9	6·7	6·9	6·2	7·9	5·4
1975/76	8·1	7·2	9·6	10·1	8·3	7·3	8·4	7·3
Whole milk	Weighted average price received by farmers for all types of sales							
1957/58	34·9	25·1	31·3	31·5	27·2	38·4	22·1	21·7
1967/68	40·2	33·7	42·9	37·7	38·1	34·4	na	28·1
1975/76	57·4	na	54·9	47·0	42·5	43·1	na	49·7

Source: Figures derived from FAO/ECE Agricultural Division of Economic Commission for Europe, *Prices of Agricultural Products and Selected Inputs in Europe and North America Series, 1957/8–1975/6.*

Table 8(a)   *Community expenditure on agricultural policies (million UA)*

	1963/4	*1968/9*	*1972*	*1977*
Community Expenditure:				
EAGGF Guarantee	50·7	1,642·6	2,700·0	6,662·4
EAGGF Guidance	9·1	94·9	376·1	289·7
EAGGF total	59·8	1,737·5	3,076·1	6,952·1

Source: EC Commission, *The Agricultural Situation in the Community 1978 Report.*

Table 8(b)   *EAGGF Guarantee section, expenditure by sector (million UA)*

	1962/3 to 1966/7		*1968/9*		*1977*	
	Amount	*%*	*Amount*	*%*	*Amount*	*%*
Cereals including rice	468·534	53·64	692·390	42·14	600·6	9·02
Dairy products	269·411	30·85	381·835	23·25	2,545·0	38·20
Pigmeat and beef and veal	33·894	3·88	50·250	3·61	442·7	6·64
Eggs and poultry meat	11·604	1·33	6·261	0·38	22·3	0·33
Fats and oils	81·093	9·28	210·496	12·81	305·0	4·58
Sugar	7·213	0·83	245·836	14·97	536·7	8·06
Fruit and vegetables	1·702	0·19	29·189	1·78	186·2	2·80
Monetary Compensatory Amounts					859·9	12·91

Source: EC Commission, *The Agricultural Situation in the Community 1978 Report.*

Table 9 *Trends in EC trade in food, beverages and tobacco*
(value in million EUA)

Year	Imports (EC 9)		Exports (EC 9)	
	Intra	Extra	Intra	Extra
1958	2,280	7,190	2,172	1,858
1959	2,430	7,312	2,372	1,764
1960	2,675	7,497	2,600	1,939
1966	4,775	9,934	4,591	2,871
1967	5,128	9,617	4,965	2,966
1968	5,659	9,246	5,516	3,179
1976	20,272	21,090	21,102	9,210
1977	23,369	24,021	23,601	10,360
1978	26,022	23,181	26,153	11,432

Source: EUROSTAT, *Monthly External Trade Bulletin*, Special Number
1958–1978.

Table 10 *EC imports and exports* (value in million EUA)*

Year	All Imports	Agricultural Imports	All Exports	Agricultural Exports
1958	20,866	7,917	20,729	2,862
1966	50,126	13,846	49,200	5,982
1967	51,605	13,491	52,729	6,409
1968	60,217	14,487	62,402	7,481
1973 Eur. 6	138,057	30,932	142,056	18,261
Eur. 9	174,650	39,766	170,817	23,082
1975	242,888	46,431	239,791	30,339
1976	305,935	57,290	291,969	35,613
1977	339,409	65,561	332,493	40,301

* 1958–73 figures for Eur. 6.
 1973–7 figures for Eur. 9.
Source: Figures derived from EUROSTAT, *Yearbook of Agricultural Statistics*,
1973–8.

Table 11   Gross Valued Added per person employed in agriculture, forestry and fishing as a per cent of Gross Domestic Product per person employed (%)

	W. Germany	France	Italy	Neth.	Belgium	Lux.	UK	Ireland	Denmark
1970	44	51	52	85	91	41	98	62	68
1971	45	51	51	79	95	42	96	62	68
1972	46	54	51	81	103	47	95	70	75
1973	46	na	56	87	106	62	101	na	95
1974	40	na	52	72	90	52	104	na	95
1975	45	na	57	77	95	63	100	na	84
1976	45	na	54	77	92	na	104	na	86

Source: MAFF, EEC Agricultural and Food Statistics, 1974, 1977, 1978.

Table 12   *Variation in prices, 1963–73* (1963 = 100)

	W. Germany	France	Italy	UK	USA
*Wheat	89·3	108·4	120·4	221·9	216·8
*Sugar beet	89·5	150·6	155·0	154·8	220·1
†Cattle	142·3	222·1	173·0	222·6	215·1
‡Butter	101·4	136·8	na	124·1	120·8

*Producer prices in all countries.
†Liveweight producer prices in Italy, the USA and Germany (1963).
Liveweight wholesale prices in the UK and Germany (1973).
Slaughterweight wholesale prices in France.
‡Producer prices in Germany (1963).
Wholesale prices in France, the UK, the USA and Germany (1973).
Source: Figures derived from FAO, *Production Yearbook*, 1964 and 1975.

Bibliography

Agostini, D., *et al.*, *Wageningen Memorandum on the Reform of the European Community's Common Agricultural Policy* (Wageningen, International Agricultural Centre, 1973).
A summary of decisions and recommendations made jointly on the future development of the CAP by economists from the nine member states of the Community. The proposals followed discussion based on papers given at the Horring Memorial Conference in 1973.

Agra Europe (Tunbridge Wells, Kent, Agra Europe (London) Ltd).
A periodical dealing with current major events in European agriculture.

Barclays Bank Ltd, *The Agricultural Common Market Beyond Transition* (London, Barclays Bank Ltd, 1977).
A simple outline of the CAP, the main policies, how the policy is worked and the problems encountered. Good introductory text.

Butterwick, M., and Neville-Rolfe, E., *Food, Farming, and the Common Market* (London, Oxford University Press, 1968).
This valuable study describes the development of agricultural policy in the UK since the war and the development of the CAP since 1958. It discusses the impact of adopting the CAP on the UK and compares production of major agricultural products in the UK and the EEC.

Butterwick, M., and Neville-Rolfe, E., *Agricultural Marketing and the EEC* (London, Hutchinson, 1972).
A clear and detailed account of the EEC arrangements for agricultural produce. It describes marketing systems in the EEC and the effect entry will have on British marketing. Although written prior to the enlargement of the Community in 1973, its thoroughness makes it still very relevant.

Central Council for Agricultural and Horticultural Co-operation, *Agricultural Co-operatives in the EEC* (London, CCAHC, 1970).
This is a comprehensive overview of the structure and organisation of agricultural and horticultural co-operatives in each member state of the EC in 1968.

Commission of the European Communities, *A Regional Policy for the Community* (Luxembourg, Office for Official Publications of the European Communities, 1969).

A memorandum setting out the objectives of a Community regional policy and establishing guidelines for financing and administration.

Commission of the European Communities, *Memorandum on the Reform of Agriculture in the European Economic Community*, Supplement to Bulletin No. 1 (Brussels, Secretariat-General of the Commission, 1969).
A study by the Commission of the problems of the CAP in 1968, covering the working of the markets as well as structural aspects of agriculture. Proposals for the reform of agriculture are outlined in detail.

Commission of the European Communities, *Stocktaking of the Common Agricultural Policy*, Bulletin Supplement 2/75 (Luxembourg, Office for Official Publications of the European Communities, 1975).
This is the most authoritative review of the policy: its past history and the extent to which it has fulfilled the Treaty of Rome objectives. It summarises the main problems and indicates possible areas of improvement.

Commission of the European Communities, 'Analysis of the stocktaking of the CAP', *Newsletter on the Common Agricultural Policy*, No. 1 (January 1976) (Brussels, Directorate-General for Press and Information, 1976).
Commentary on the document 'Stocktaking of the Common Agricultural Policy'.

Commission of the European Communities, *The Agricultural Situation in the Community*, annual reports 1975–8 (Luxembourg, Office for Official Publications of the European Communities, 1976, 1977, 1978).
This series of annual reports gives a comprehensive overview and analysis of the agricultural situation in the Community. They are an excellent source of reference for up-to-date statistics and contemporary issues in all member states of the European Community.

Commission of the European Communities, *Community Regional Policy (New Guidelines)*, Bulletin Supplement 2/77 (Luxembourg, Office for Official Publications of the European Communities, 1977).
A revision of the recommendations and objectives for a regional policy for the Community.

Commission of the European Communities, *The Common Agricultural Policy* (Brussels, Commission of European Communities, 1977).
A clear, comprehensive description of the provisions of the CAP, including the regulations governing individual commodities and the structural directives, and how the policy works in practice.

Commission of the European Communities, *The Agricultural Policy of the European Community* (2nd edn) (Luxembourg, Office for Official Publications of the European Communities, 1979).
An excellent résumé of the Community's agricultural policy, more analytical than descriptive, to be read after a knowledge of the basic policy provisions has been grasped.

Corden, W. M., *Monetary Union: Main Issues Facing the European Community*, International Issues No. 2 (London, Trade Policy Research Centre, 1976).

A critical discussion of the possibilities for monetary union within the EC.

Falk, R., *Agricultural Co-operatives in an Enlarged EEC*, Conference Paper No. 8 (Newcastle-upon-Tyne, Agricultural Adjustment Unit, University of Newcastle, 1972).
This report compares the development of co-operatives on the Continent with their development in the UK, assessing the differences in their financial and legal situations, their relative strength as a commercial force and the future contribution of co-operatives to the enlarged Community.

Federal Trust Conference Report, *Current Agricultural Proposals for Europe* (London, Federal Trust, 1970).
Written in 1970, this report deals with the Mansholt Plan and the Vedel Report, and their proposals for reforming agricultural policy in Europe.

Fennell, R., *The Common Agricultural Policy: A Synthesis of Opinion*, CAES Report No. 1 (Ashford, Kent, Centre for European Agricultural Studies, Wye College, 1973).
The CAP has been the subject of much comment and criticism. This study collates some of the considerable literature written in the late 1960s and early 1970s.

Food and Agriculture Organisation of the United Nations, *Production Yearbook* (Rome, FAO).
Food and Agriculture Organisation of the United Nations, *Trade Yearbook* (Rome FAO).
Annual publications of world production and trade statistics.

Harris, S., *The World Commodity Scene and the Common Agricultural Policy*, Occasional Paper No. 1 (Ashford, Kent, Centre for European Agricultural Studies, Wye College, 1975).
A discussion of the CAP, changes in its domestic policies in 1975 and changes in its external relations, especially with developing countries.

Heidhues, T., Josling, T. E., Ritson, C., and Tangermann, S., *Common Prices and Europe's Farm Policy*, Thames Essay No. 14 (London, Trade Policy Research Centre, 1978).
This report analyses the pricing system of the CAP as an example of the problems faced in economic policy-making. It examines the evolution of the CAP, identifying the lack of a common European currency as one of the major drawbacks.

Herlitska, A., Malve, P., and Winegarten, A., 'European agricultural policy', in *US Agriculture in a World Context*, ed. D. G. Johnson and J. A. Schnittker (New York, Praeger, 1974).
This paper gives a concise outline of the evolution of the CAP, looking particularly at the trade, income and social aspects of the policy. It looks briefly at the future for the CAP.

Irving, R. W., and Fearn, H. A., *Green Money and the Common Agricultural Policy*, Occasional Paper No. 2 (Ashford, Kent, Centre for European Agricultural Studies, Wye College, 1975).
A comprehensive report contributing to the debate about 'green money' and its

effects. The study outlines the development and role of green money against the background of common pricing, the 'cornerstone' of the CAP.

Josling, T. E., and Hamway, D., 'Income transfer effects of the Common Agricultural Policy', in *Agriculture and the State*, ed. B. Davey, T. E. Josling and A. McFarquhar (London, Macmillan for the Trade Policy Research Centre, 1976).
The income distribution effects and inter-country transfers involved in Britain's entry to the EC are analysed.

Josling, T. E., and Harris, S., 'Europe's green money', *The Three Banks Review*, no. 109 (March 1976), pp. 57–72.
A concise article tracing the inception, development and effects of the green money system.

Knox, F., *The Common Market and World Agriculture: Trade Patterns in Temperate-Zone Foodstuffs* (New York, Praeger, 1972).
Although written prior to Britain joining the EC, this study examines very well the effect of British entry on the structure of international trade in agricultural goods. Chapter 2 is especially good in its outline of the development and trade effects of the CAP.

Lane, P., *et al.*, 'The Common Agricultural Policy in question', *World Agriculture*, vol. 25, nos 2–3 (1976), pp. 3–27.
A concise review of the working of the CAP, aiming, in the light of the criticisms levelled at it, to reflect the on-going discussions on its present position.

Mackel, C. J., *The Development, Role and Effects of Green Money in a Period of Economic Instability*, Bulletin No. 13 (Aberdeen, North of Scotland College of Agriculture, 1977).
A clear, informative study on the reasons for and development of green money and a good guide to the effects of monetary compensatory amounts.

MacKerron, G., and Rush, H. J., 'Agriculture in the EEC – taking stock', *Food Policy*, vol. 1, no. 4 (1976), pp. 286–300.
An assessment of the Community's 'Stocktaking of the CAP'.

MacLennan, R., 'Food prices and the Common Agricultural Policy' *The Three Banks Review*, no. 119 (September 1978), pp. 58–71.
A concise article dealing with the latest developments in the green money system. It concentrates on the impact of the CAP on UK food prices but in a wider context argues that the objectives of the Treaty of Rome cannot be met at current degrees of disparity in economic performance among member states.

Marsh, J. S., *rapporteur*, *European Agriculture in an Uncertain World*, The Atlantic Papers No. 1 (Paris, Atlantic Institute for International Affairs, 1975).
A consideration of European agricultural policy by a panel of economists. On the basis of an examination of the CAP and the environment in which it operates, suggestions have been made for a change of emphasis within the policy.

Marsh, J. S., 'European agricultural policy: a federalist solution', *New Europe* (Winter 1977), pp. 27–40.

The author examines the reasons behind the CAP's present failure and offers suggestions for reform of the CAP based on a regional approach: that the Community abandon the attempt to make uniform, centralised decisions and admit national differences but yet maintain a system of trade that would progress towards a more truly common market.

Marsh, J. S., *UK Agricultural Policy within the European Community*, CAS Paper 1 (Reading, Centre for Agricultural Strategy, 1977).
UK agricultural policy is constrained by membership of the EC and the terms of the CAP. This report suggests a UK strategy to promote CAP reform to allow greater scope for UK agricultural policy.

Marsh, J. S., and Ritson, C., *Agricultural Policy and the Common Market*, European Series No. 16 (London, Chatham House and Political and Economic Planning, 1971).
A study which examines the nature and development of the CAP. By analysing the theoretical background against which policy decisions are taken, it explores the sources of the difficulties of policy formulation in the EC.

Meat and Livestock Commission, Economic Information Service, *The Common Agricultural Policy: Beef and Veal* (Milton Keynes, MLC, 1978).
Meat and Livestock Commission, Economic Information Service, *The Common Agricultural Policy: Pigmeat* (Milton Keynes, MLC, 1978).
Meat and Livestock Commission, Economic Information Service, *Green Money and the Meat and Livestock Industry* (rev. edn) (Milton Keynes, MLC, 1978).
These publications outline in detail the EC regimes for beef and veal and pigmeat and the effects of the agri-monetary situation on livestock trade, particularly the difficulties associated with monetary compensatory amounts.

Meat and Livestock Commission, Economic Information Service, *EEC Statistics*, vols 1–3 (Milton Keynes, MLC, 1979).
Livestock numbers, production, consumption and prices for all member states of the EEC.

Milk Marketing Board (Economics Division), *EEC Dairy Facts and Figures, 1978* (Thames Ditton, Surrey, MMB, 1979).
Statistics on EEC dairy herd numbers, milk production, etc.

Ministry of Agriculture, Fisheries and Food, *Agricultural Statistics in an Enlarged EEC* (London, MAFF, 1976).
Basic data relating to agriculture and food in the EC. The figures generally relate to one or two recent years.

National Farmers' Unions of England and Wales and of Scotland and the Ulster Farmers' Union, *EEC Policies for Improving Farm Structures* (London, NFU, 1974).
A discussion on the criticisms levelled at the CAP, in particular its 'failure' to achieve the aim of improving farm incomes, by consideration of the existing commodity regulations and the scope for the introduction of more selective income aids.

National Farmers' Unions of England and Wales and of Scotland and the Ulster Farmers' Union, *EEC Policies for Improving Farm Structures* (London, NFU, 1974).
This booklet deals in detail with Community measures to improve the structure of agricultural production and marketing. A very good guide to structural policy provisions up to 1974, especially the three directives of the early 1970s.

Organisation for Economic Co-operation and Development, *Agricultural Statistics 1955–1968* (Paris, OECD, 1969).
This volume contains most of the main agricultural statistics available for the years 1955 to 1968.

Organisation for Economic Co-operation and Development, *Structural Reform Measures in Agriculture* (Paris, OECD, 1972).
A report on the structural reform measures taken by various countries in recent years. It examines the problems and the degree of success achieved in solving them and indicates what type of action might be best in the future. It concentrates particularly on 'external' rationalisation or measures to improve the 'shape' or increase the physical size of farms rather than on intensification of existing production systems.

Organisation for Economic Co-operation and Development, *Agricultural Policy of the European Economic Community* (Paris, OECD, 1974).
Fairly comprehensive and clear outline of price and structural policies of EEC agriculture up to 1974.

Priebe, H., Bergman, D., and Horring, J., *Fields of Conflict in European Farm Policy*, Agricultural Trade Paper No. 3 (London, Trade Policy Research Centre, 1972).
This report contains three essays on, respectively, the German, French and Dutch viewpoints on the CAP. It is invaluable as a guide to the opinions held in other European countries.

Statistical Office of the European Communities, *Agricultural Statistics* (Luxembourg).
Statistical Office of the European Communities, *Yearbook of Agricultural Statistics* (Luxembourg).
Regular official publications on production, income, capital, etc., in agriculture in each member state of the EC.

United Nations Economic Commission for Europe and Food and Agriculture Organisation, *Prices of Agricultural Products and Selected Inputs in Europe and North America*, annual FAO/ECE price reviews (New York, United Nations).
Data on farm-gate prices and an 'average' price for agricultural products in every European and North American country.

Weinschenck, G., 'Basic alternatives of the future agricultural policy in the European Common Market', *Journal of Agricultural Economics*, vol. 26 (1975), pp. 145–158.
The paper examines the alternatives for the future development of the CAP. It argues that without regional differentiation there is little scope for a reorganisation of policy instruments.

Index